The Surface Scientist's Guide
to Organometallic Chemistry

The Surface Scientist's Guide to Organometallic Chemistry

Mark R. Albert

and

John T. Yates, Jr.

American Chemical Society

Washington, DC   1987

Library of Congress Cataloging-in-Publication Data

Albert, Mark R., 1954–
 The surface scientist's guide to organometallic
chemistry.

  Bibliography: p.
  Includes index.

  1. Organometallic chemistry. 2. Surface chemistry.

  I. Yates, John T., 1935–   . II. Title.

QD411.A424   1986   547'.05   86–25937
ISBN 0–8412–1003–9

547.25

287

About the Authors

Mark R. Albert is a research associate at Princeton University. He has been involved in the fields of organometallic chemistry and surface science since he was an undergraduate at the University of Maryland. There he studied organometallic chemistry and undertook honors research in the area of electron probe microanalysis of asbestos fibers. At the University of Maryland he was awarded the Noxell Corporation Scholarship Award. After graduation from Maryland, he spent a summer at the Eastman Kodak Corporation where he synthesized and studied transition metal complexes. In his graduate work at the University of Pennsylvania, he continued his study of organometallic chemistry and learned a variety of synthetic methods that he applied to an ultraviolet and X-ray photoelectron spectroscopy study of alkyne hexacarbonyl cobalt complexes. He also worked closely with the surface science group in the Department of Physics and studied the chemisorption of hydrocarbons on single-crystal surfaces, also with photoemission. This research gave him the opportunity to work at the Synchrotron Radiation Center at the University of Wisconsin. After receiving his Ph.D., he worked for the Dow Corning Corporation in the area of electron spectroscopy for chemical analysis and Auger spectroscopy and then went to the Surface Science Center at the University of Pittsburgh, where he coauthored this book. His research at the University of Pittsburgh involved the study of small-molecule chemisorption on supported metals using transmission infrared spectroscopy. He is presently at Princeton University, where he is studying chemisorption phenomena with high-resolution electron energy loss spectroscopy (HREELS).

REVIEW

JOHN T. YATES, JR.,
received his B.S. degree from Juniata College and
his Ph.D. in physical chemistry from the Massachu-
setts Institute of Technology. Following a three-year
term as Assistant Professor at Antioch College, he
joined the National Bureau of Standards, first as a
National Research Council Postdoctoral Research
Fellow and then, from 1965 until 1982, as a member
of its scientific staff. His research in the fields of
surface chemistry and physics, including both the
structure and spectroscopy of surface species, the
dynamics of surface processes, and the development
of new methods for research in surface chemistry,
has put him in the forefront of an exciting and rapidly growing field of
science. He is the author of more than 200 publications. Yates was Senior
Visiting Scholar at the University of East Anglia, Norwich, in 1970-71, and
Sherman Fairchild Distinguished Scholar at California Institute of Technology in 1977-78. He received the Silver Medal of the U.S. Department of
Commerce in 1973, the Stratton Award for Distinguished Research at the
National Bureau of Standards in 1978, and the Gold Medal—the highest
award of the U.S. Department of Commerce—in 1981. In 1986, he received
the Kendall Award in Colloid or Surface Chemistry from the American
Chemical Society.

Yates joined the University of Pittsburgh in 1982 as the first R. K. Mellon
Professor of Chemistry and as the first Director of the University of
Pittsburgh Surface Science Center. Here, working with students and
postdoctoral staff, his influence actively extends over a wide range of
research projects in surface science. He is also active in undergraduate and
graduate teaching. In addition, he maintains close relationships with many
surface science research programs in academic, government, and indus-
trial research laboratories throughout the world and serves on the editorial
boards of five journals and two book series in surface science and
catalysis.

Contents

Preface

WITHIN THE FIELD OF SURFACE SCIENCE, concepts from the fields of organometallic and coordination chemistry are becoming increasingly relevant as a means of understanding chemisorption systems. This situation has occurred because the two fields work with the same metals and the same ligands that bond to these metals, and very often they pursue similar research issues. But the two fields employ a different language and ways of gaining experimental information. These differences are detrimental to surface science research because many insights into ligand structure and bonding that are common to organometallic chemistry are hidden from the surface science community. This book was written, therefore, as a means of communicating to the surface science community the aspects of organometallic and coordination chemistry that are relevant to the research issues currently being pursued by surface scientists.

The many ligands commonly used by the coordination chemist are all good candidates for study on surfaces, and it would not be surprising for each ligand to enhance the understanding of surface processes to the extent already accomplished by CO chemisorption research. The application of theoretical calculations useful within coordination chemistry research to surface science research seems to be a natural extension of the experimental organometallic chemistry research that has already contributed to the understanding of chemisorption systems. However, this book is only a beginning in surface science's full exploitation of organometallic and coordination chemistry. It emphasizes issues concerning characterization, bonding, and preparation of ligand bonding modes. Issues related to heterogeneous catalytic chemistry are not directly addressed. Most likely, organometallic and coordination chemistry will provide a similar degree of insight into these issues, where ligand–ligand interactions become important.

But the exploitation of coordination chemistry by surface scientists does not come without its challenges. No indexes or computer searches can painlessly extract from the coordination chemistry literature analogies or new research ideas that will interest the surface scientist. This book can only be a beginning in this regard. The only real solution to this problem is through diligence in the study of the coordination chemistry literature and by following the new developments in coordination chemistry as closely as new surface science developments are followed. Some of the journals and monograph series worth following include *Journal of the American Chemical Society, Chemical Reviews, Journal of Coordination Chemistry, Inorganic Chemistry, Organometallics, Journal of Organometallic Chemistry, JCS Dal-*

ton Transactions, Coordination Chemistry Reviews, Advances in Organometallic Chemistry, Progress in Inorganic Chemistry, Advances in Inorganic Chemistry, and Radiochemistry. Most of the coordination chemistry references in this book are from one of these journals.

At this stage there appears to be no limit to the number of new ideas that surface science can draw from coordination chemistry research. However, coordination chemistry is an active, ongoing field of research that will continue to contribute to the further understanding of surface chemisorption systems. This kind of interaction is dependent only upon the communication that exists between surface scientists and coordination chemists. This book is only the beginning of this challenging and inevitably beneficial interaction.

Acknowledgments

We wish to acknowledge support of this work by the 3M Science Research Laboratory and the Corporate Research Laboratory. A special note of thanks is given to Kurt Kolasinski, who drew all of the molecular structures in this book; the excellent quality of his work has improved the clarity of this review enormously. We especially acknowledge the encouragement and support of Allen Siedle of the 3M Corporation during the writing of this book. Another special note of thanks is given to Michael Bozack, who provided extensive editing of the text and improved the book's readability. Thanks are also offered to Rex E. Shepherd, E. Ward Plummer, Miles J. Dresser, Larry G. Sneddon, John E. Crowell, and Valerie Shalin, who read and offered helpful comments about the manuscript prior to publication. Appreciation is also expressed to Neil R. Avery who provided several of his publications prior to their publication. Many thanks are also offered to Phyllis Roy, Mary Ann Stevwing, Cathy Veneskey, and Nancy Caldiera for their excellent typing of the many drafts of the manuscript.

MARK R. ALBERT
Princeton University
Princeton, NJ 08540

JOHN T. YATES, JR.
University of Pittsburgh
Pittsburgh, PA 15260

Dedication

THE FIRST DRAFT OF THIS BOOK was about half-finished when the world of organometallic and inorganic chemistry was saddened by the death of Earl Muetterties in Berkeley. Because of his significant contributions to organometallic chemistry and surface science, we dedicate this book to his memory.

His perception of the possible relationship between organometallic chemistry and chemisorption and catalysis on surfaces has been a driving force among many surface chemists. This book would not now exist if Earl Muetterties had not set his theory of the metal cluster–surface analogy down on paper. Already numerous works based on the cluster–surface analogy exist, and the many more to follow will undoubtedly uncover new insights into chemisorption and surface reactivity.

Earl Muetterties worked for his students and for the field of surface chemistry in ways that his friends and colleagues will always remember. We received a personal letter from him, mailed 14 days before his death, containing recent preprints from his group. One of us (M.A.), who did most of the writing and thinking about this book, had hoped that Earl Muetterties's friendly and perceptive criticism might have played a role in making this a better book. The other one of us (J.Y.) had hoped that, through Earl Muetterties's work, the field of surface science would undergo a distinctive revolution from a physics-dominated, technique-oriented subject to a more chemically oriented field, filled with molecular orbitals, interesting structures, interesting chemical concepts, and new classes of surface reactions.

It remains for all of us to carry on in the tradition and direction first charted by Earl Muetterties. We hope that this book in some small way may be a tribute to Earl Muetterties, his work, and his students.

MARK R. ALBERT
JOHN T. YATES, JR.

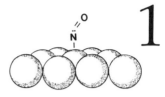

Introduction

THE INCREASED UNDERSTANDING OF THE INTERACTION of gases with transition metal surfaces has been accompanied by a growing awareness that many gas–surface interactions have analogues in the literature of organometallic and coordination chemistry. This analogy, also known as the cluster–surface analogy, was first proposed by Muetterties (1, 2), and he and others have since developed this idea in more recent works (3-6). Opposing opinions have also been expressed (7). Since Muetterties's pioneering efforts, a number of workers studying various chemisorption systems have attempted to interpret data in light of analogous transition metal–ligand interactions. These include studies comparing photoemission spectra (8-12) and, more commonly, works comparing the frequencies of ligand vibrations with those of the corresponding surface-bound species (13-19). The general conclusion of these studies has been that, qualitatively, the nature of the gas–surface interaction is similar to the metal–ligand interaction. In addition to the direct experimental studies, Muetterties (3, 6) also hypothesized that similarities in reactivity should be expected in both metal clusters and surfaces, and he found many correlations between surface-bound molecules and ligands in clusters.

In spite of this growing relevance of coordination chemistry to chemisorption, there is as yet no means to easily extract information from the coordination chemistry literature that may be relevant to surface studies. The voluminous literature of coordination chemistry is composed largely of information that is not pertinent to the study of chemisorption systems. This extra information includes methods of preparing coordination compounds, the characterization of large complexes that have a variety of coordinated ligands, and the reactions of these complexes. Although these areas of research are quite challenging, the resulting body of information only obscures the issues that are important to surface science research. These relevant issues include the nature of the bonding interaction of a given ligand with a metal center, the number of possible ways that a given ligand may bond to a metal center or cluster, and how the bonding mode of a given ligand can be characterized spectroscopically. Coordination chemists have studied these areas in detail, but isolating this information in the literature can at times be a very tedious task.

This book is intended to be a means of simplifying this process by reviewing the numerous ligands commonly used in coordination chemistry and discussing the aspects of each that might aid in the understanding of how that ligand may bond to a surface. A less extensive review of organometallic ligands has already appeared (*20*). In addition, the nature of coordination sites in transition metal complexes and clusters is explored in the hope that the insights gained from understanding these coordination sites will add to the understanding of bonding sites on surfaces. Also, it is hoped that this book will generate new ideas for chemisorption studies that may not have been considered previously. Those interested in gaining an exposure to the basics of surface science research should refer to one of the many works on the subject (*3, 21*).

Although Muetterties (*6*) has limited his analogy of surface bonding to organometallic cluster compounds, this review will encompass the bonding of ligands to any metal center, be it in clusters or the much more common monometal center. The research just mentioned and the further comparisons of monometal complexes and chemisorbed species to be discussed later show that a ligand on a surface usually bonds in a manner similar to that on monometal centers, clusters, and surfaces. Therefore, it seems reasonable to extend the cluster–surface hypothesis to any ligand that bonds to a metal center in coordination complexes.

The issue of greatest concern regarding the analogy of organometallic complex bonding to surface chemisorption lies in the fact that although metal complexes and small clusters can be described by discrete orbitals, surfaces are usually described by band structure (*22, 23*). Some believe that this difference invalidates the analogy to be considered in this book. However, the final chapter of this book presents recent calculations that appear to effectively bridge this gap and provide a method for extending the experimental and theoretical understanding of metal–ligand structure and bonding developed in organometallic chemistry to surface chemisorption (*24, 25*).

This review is structured in the following way: Following this introduction, a short discussion of some fundamental concepts in coordination chemistry is presented to provide a basis in understanding for those who may be unfamiliar with the fundamentals of organometallic chemistry. Next, the various ligands used in coordination chemistry are discussed in terms of bonding modes, orbital overlaps, molecular orbital diagrams when available, and methods of characterizing each ligand's individual bonding modes.

Following this largest chapter, the molecular orbital structures of metal complex and cluster bonding sites are presented. The final chapter of this book discusses the theoretical treatment that can bridge the gap between orbital arguments and issues of surface band structure.

The characterization methods discussed in Chapter 3 are limited almost exclusively to vibrational spectroscopy* because it is the most common technique used in synthetic coordination chemistry and is the only technique readily available to both coordination chemistry and surface science. Techniques such as nuclear magnetic resonance (NMR) spectroscopy and X-ray crystallography, which are powerful characterization tools for the chemist, are not as yet generally applicable to surface science studies on single-crystal surfaces, even though some reports of NMR studies of chemisorbed species have appeared in the literature (26-29), as well as in a literature review (30). Low-energy electron diffraction (LEED) can approximate the high accuracy of X-ray crystallography, and many chemisorption systems have been studied in detail by using dynamical LEED calculations (21, 31, 32). However, as yet, there is little overlap between the surface and metal complex systems studied by the two techniques. This situation complicates comparisons of organometallic ligands and surface adsorbates through structural similarities. Surface extended absorption fine structure (SEXAFS) (33) is a more exact method of determining bond lengths of chemisorbed species. However, it is still a relatively new technique, and relatively few surface systems have been studied with SEXAFS.

Conversely, ultraviolet photoemission spectroscopy (UPS), a significant surface analysis tool, is of limited use to the coordination chemist because relatively few of the known transition metal complexes have the high symmetries required to resolve all the energy levels of a single ligand. Many complexes have a variety of ligands that usually result in complicated UPS spectra in which the peaks of the various ligands overlap. In addition, most organometallic complexes are not sufficiently volatile, or decompose with heating and thus could not be analyzed at high resolution in the gas phase. Electron spectroscopy for chemical analysis (ESCA) can be used to study both surface adsorbates and coordination complexes; however, few comparisons have been made between coordination compounds and surface species with ESCA (34-36). In addition, there are few ESCA studies of a given ligand in a series of electronic environments in metal complexes—studies that can correlate binding energies and peak splittings with

---

*Several works have extensive compilations of characteristic vibrational data of organometallic compounds. Infrared and Raman Spectra of Inorganic and Coordination Compounds, Third Edition (John Wiley and Sons: New York), by K. Nakamoto, is one of the more complete works. Physical Methods in Chemistry (W. B. Saunders: Philadelphia), by R. S. Drago, also contains extensive compilations of vibrational data as well as a good discussion of the theoretical aspects of vibrational spectroscopy. Other good compilations include Vibrational Spectra of Organometallic Compounds (John Wiley and Sons: New York), by F. Maslowsky, and Infrared Spectra of Inorganic Compounds (Academic: New York), by R. A. Nyquist and R. O. Kagel.

structure (*34–36*). The ESCA studies that relate to characterizing a ligand's bonding modes, however, are mentioned in this book.

Because UPS apparently has a greater applicability to chemisorbed species than to metal complexes, it is possible, as an added benefit of surface studies, that insight into the bonding of a ligand in a complex could be gained by studying the UPS spectra of that ligand adsorbed on a transition metal surface. The absence of the other ligand moieties that compose a metal complex could eliminate the problem of the overlapping peaks for the ligand under study.

It is impossible in the limited space of this book to discuss every single small molecule used as a ligand. This problem, therefore, has been mitigated by organizing the various types of ligands into a series of 13 classes that share similar bonding modes. The boundaries of the classes may be fuzzy and some overlap may occur between the classes, but, on the whole, the classes seem to be generally unique. The only class not represented in this book is that of the macrocycles, a series of large planar molecules containing a central hole large enough to accommodate a coordinated transition metal atom. These are excluded because macrocycles will probably be unable to chemisorb on flat metal surfaces by using the same mode of coordination that is used in the complexes.

For a book, such as this one, that presents one field of research compared to another, a problem with vocabulary inevitably occurs because commonly used words in one field are often unknown in another. To alleviate this problem, a fairly extensive glossary is included in this book, starting on page 191. In addition, terms unique to organometallic chemistry are avoided where possible. The analogous term in surface science is used if possible, or if that is not possible, the chemistry term is defined when it is first used. In some cases, the standard nomenclature of molecular structure is avoided so that more descriptive references can be made to ligands in metal complexes. If the reader does encounter a strange word, and if its definition is not obvious from the context or the glossary, referral to almost any inorganic chemistry text should be sufficient to resolve the confusion. One term that is worth noting here is the "R" that commonly appears in molecular structures. It is simply a variable indicating any one of a great variety of molecular structures. The use of R in a structure indicates that that part of the molecule has no bearing on the chemistry being discussed. Another term worth mentioning here is *sterics,* which essentially refers to the limitations in bonding and reactivity caused by the fact that only one atom can occupy a given space at a given time.

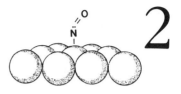

2

Basic Coordination Chemistry

PRIOR TO IN-DEPTH DISCUSSION OF THE LIGANDS used in coordination chemistry, some of the fundamental concepts in coordination chemistry that can provide a general picture of the nature of the bonding in complexes should be discussed. Although the structure and bonding of transition metal complexes are intensively studied by numerous experimental and theoretical methods, relatively few concepts need to be discussed to provide a basic understanding of coordination chemistry, especially as to how it relates to the chemisorption of gases on surfaces. These concepts are discussed in this chapter.

2.1   The Octet and 18-Electron Rules

The fundamental driving force for all chemical bonding is stated in what is called the *octet rule* (37), that is, all main group atoms can achieve a maximum stability by acquiring an outer, principal quantum shell that contains eight electrons and thus is filled. An atom can attain this closed valence shell in a number of ways. It can be done by ionization, that is, by gaining or losing electrons. The chlorine atom ionizes when it acquires its eighth valence electron and becomes the chloride ion, which has a filled valence shell; the sodium atom ionizes when it gives up its single valence electron to form the sodium cation, which has a closed valence shell composed of the atom's next lower principal shell. More commonly, however, atoms attain a filled valence shell by sharing valence electrons with other atoms through the formation of molecular orbitals with those atoms, that is, covalent bonding. For example, when a nitrogen atom bonds to another nitrogen atom to form the N_2 dimer, each nitrogen atom is using five of its own electrons and three shared electrons from the other nitrogen

1003-9/87/0005$06.00/1 © 1987 American Chemical Society

atom to gain a filled valence shell with eight electrons. A third method of attaining a closed valence shell, the method of greatest importance to coordination chemistry, involves one atom sharing an electron pair that is localized on another atom. For example, the boron atom in BCl_3, which has only six valence electrons, can acquire a closed shell of eight electrons by sharing the lone electron pair on the nitrogen atom in an ammonia molecule to form the adduct $Cl_3B \leftarrow NH_3$. This type of bonding is called coordinate covalent or dative bonding.

When transition metal atoms become involved in the bonding, that is, atoms with partially filled d orbitals, the octet rule is extended and becomes the 18-electron rule to account for the additional 10 electrons required to fill the d subshell (*38*). Therefore, a chromium atom with six valence electrons (one in the s shell and five in the d shell) will bond with six lone-pair donors such as CO to acquire 12 additional valence electrons and, thus, a filled valence shell of 18 electrons. For the chromium atom as well as for all the transition metals, sharing lone pairs of ligands is their only alternative for acquiring a closed shell of 18 electrons because they do not have enough electrons themselves to form only covalent bonds as a means of acquiring 18 electrons. This prerequisite leads to a transition metal chemistry that is unique compared with the chemistry of the main group elements.

This 18-electron rule can be extended to cluster species by simply realizing that metal–metal bonds are not usually dative in nature, but are instead covalent. For a given metal center in a cluster, therefore, each of its metal–metal bonds counts as one additonal electron for the 18-electron count. This method is a very common way for a metal atom to close out its 18-electron shell when the metal atom contributes an odd number of electrons to its valence shell. As metal clusters get large, however, this formalism tends to break down; therefore, more elaborate molecular orbital theories are required to account for cluster bonding (*39*). Because the formalism of the metal–metal covalent bond applies to the large majority of multimetal complexes, it is generally an adequate concept for understanding most cluster systems.

From this discussion, therefore, it is possible to generally predict that only molecular species with electron pairs available for donation will make good ligands in coordination complexes. These electron pairs should be located on the molecule in a geometry that will allow them to interact with the transition metal atom. Good electron pair donors generally have an electron pair in an orbital on the ligand that is oriented away from the remaining atoms in the ligand. Electron pairs that are located between other atoms, for example, σ-type bond pairs, are not generally good electron-pair donors. The orbital containing the electron pair should also be of a higher energy than the empty d orbitals of the transition metal atom to provide an impetus for the electron donation to occur.

2.2   Theories of Orbital Overlap

For molecules to carry out this sharing of electrons, an appropriate overlap of the orbitals of each atom must occur. Chemists describe this overlap in one of two different ways depending upon the specific chemical application they are studying; although chemists have different approaches to describing orbital overlap, each approach makes important contributions to describing and understanding bonding.

The orbital overlaps that lead to this bonding between atoms occur between only the highest principal shells, that is, valence shell orbitals of each atom undergoing bonding. These valence atomic orbitals have shapes determined by their orbital angular momentum quantum number, l. When $l = 0$, the orbital is spherical and has no nodal planes and is labeled s (Figure 2.1a). When $l = 1$, the orbital is labeled p and is shaped much like a barbell (Figure 2.1b); it has one nodal surface between the two halves or lobes of the orbital. This nodal surface causes the two lobes to have opposite signs. Each principal shell above $n = 1$ contains three orthogonal p orbitals, each located along one of the Cartesian coordinate axes x, y, and z. When $l = 2$, the orbital is labeled d and has a rosette shape (Figure 2.1c). The two nodal surfaces of the d orbitals lead to opposite mathematical signs for adjacent lobes. Each principal shell above $n = 2$ contains five orthogonal d orbitals (40). The sign of each lobe refers to the sign of the mathematical function describing the lobe and not to that of electronic change.

Common to both descriptions of orbital overlap is the nomenclature of the bonds that are formed from the orbital overlap (37). Sigma (σ) bonds are formed when the electron density of the bond lies directly between the nuclei of the bound atoms. Pi (π) bonds are formed when the electron density of the bond lies off of the internuclear axis, lying only in either the xz plane or the yz plane if the internuclear axis is the z axis. The internuclear axis will contain a nodal surface of the orbital, and the two resulting lobes will correspond to wave functions of opposite sign. Therefore, when CO bonding is discussed, the donation of the 5σ electron density from the CO ligand to the metal center is σ in nature because the bonding electron density lies between the metal atom and the carbon atom. The d to π^* back donation, however, is a π-type bond because the bonding electron density lies off the internuclear axis in either the xz or yz plane and has lobes corresponding to wave functions of opposite sign.

2.2.1   Valence Bond Theory

The more qualitative of the two bonding descriptions is valence bond theory (40). It uses hybrid orbitals, which are combinations of s, p, and d atomic orbitals, to rationalize molecular structure in terms of orbital overlap. The hybrid orbitals, as a rule, bond in only the σ sense; unrehybridized p

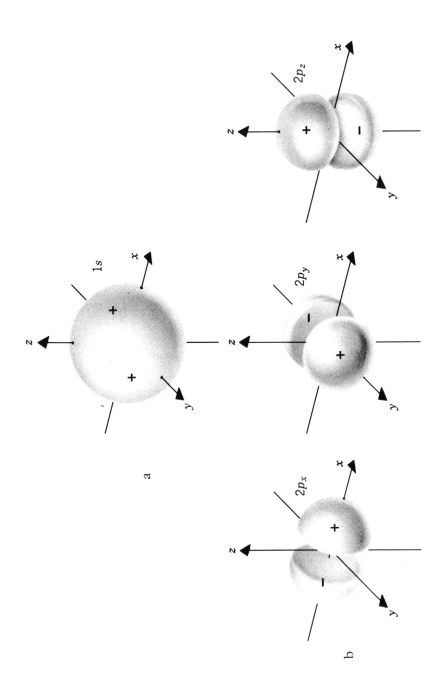

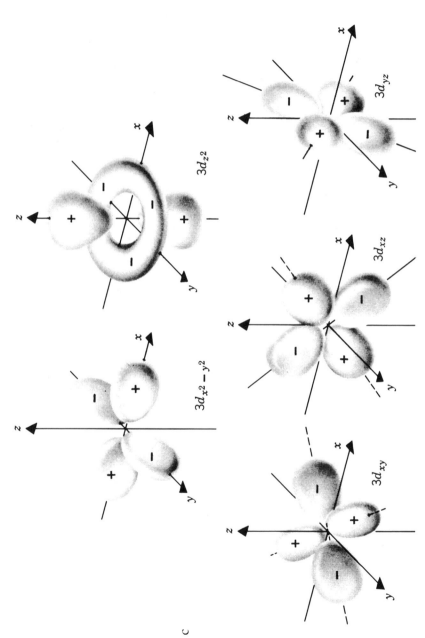

Figure 2.1. Shapes and spatial orientations of the *s*, *p*, and *d* atomic orbitals. (a) the spherical *s* orbital, (b) the three barbell-shaped *p* orbitals, and (c) the five rosette-shaped *d* orbitals. (Reproduced with permission from reference 44. Copyright 1976 Wiley.)

orbitals generally do the π-type bonding. Thus, for a structure such as methane (Structure **2.1a**), the tetrahedral geometry can only be rationalized by mixing the one s and three p valence orbitals of the carbon atom to form four hybrid sp^3 orbitals that can point to the apices of a tetrahedron and form four covalent σ bonds with four hydrogen atoms. For ethylene (Structure **2.1b**), two p orbitals mix with the s orbital to form the three sp^2 hybrids for σ bonding, whereas the unrehybridized p orbitals on either carbon atom can form the π bond between the two carbon atoms. For acetylene (Structure **2.1c**), only one p orbital mixes with the s level to generate two sp hybrid orbitals, and two p orbitals are left free to form the two acetylene π bonds. When the hybrids are formed, they always repel each other to their fullest extent and thus predict the bond angles actually found in the molecular species. Although these descriptions are not related to the reality of molecular orbital theory, they are used commonly in synthetic chemistry and provide highly reliable qualitative descriptions of the structure of essentially any molecule.

Structures **2.1a-2.1c**. Valence bond structures of (a) methane, (b) ethylene, and (c) acetylene. Each illustrates the structure resulting from sp³, sp², and sp hybridizations, respectively, on the carbon atom.

In valence bond theory, all of the bonds are defined as localized, and molecular structures can be represented by the stick figures exemplified in Structures **2.1a-2.1c**. Each stick represents one electron pair that is causing either a σ or π bonding interaction between the atoms it joins. Electron pairs that do not participate in bonds are represented by a dash or more commonly by two dots close to the atom on which they are associated. For some molecular structures, however, the sticks can be drawn into the structure in more than one way. For these cases, all of the individual structures are considered correct valence bond structures and are called resonance structures of the given molecule. The real electronic structure of the molecule is thought to be an average of the various resonance structures and is essentially a case for which the valence bond picture breaks down. The existence of resonance structures, however, is indicative of a molecule that is stabilized relative to a similar species with no resonance forms.

2.2.2 Molecular Orbital Theory

The more complex yet more quantitatively and physically accurate picture of bonding is molecular orbital theory (*41*), which describes the bonding in a molecule by considering the interaction of all of the orbitals in a delocalized fashion. One aspect of molecular orbital theory is the conservation of orbitals when a bond is formed. As shown in Figure 2.2, which describes the bonding in a molecular orbital sense for the hydrogen molecule, when two half-filled atomic orbitals overlap, two new molecular orbitals are generated. The lower energy orbital is a bonding orbital, a σ-type orbital, whereas the higher energy orbital is the antibonding orbital, labeled σ^*. The bonding arises because only the bonding orbital is filled with electrons. If the hydrogen atoms in the figure were replaced by helium atoms, each containing two electrons, then the antibonding level would also be filled, and no net bonding would occur.

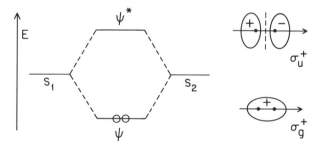

Figure 2.2. Molecular orbital picture of dihydrogen. ψ and ψ^* are the bonding and antibonding molecular orbitals, respectively, resulting from the overlap of the *s* orbitals on each atom. Note that only the bonding orbital is filled. Approximate shapes of the two orbitals are shown on the right. (Reproduced with permission from reference 45. Copyright 1977 W. B. Saunders.)

This fundamental picture can be generally applied to any bonding situation. If the atomic orbital energy levels differ, the same picture will be obtained, except that the lower energy bonding level will resemble the lower energy orbital more than the higher energy one; the reverse will be true for the antibonding level. A similar situation will arise if one of the atomic orbital levels is filled and the other empty, if the levels are forming π bonds instead of σ bonds, or if the precursor levels themselves are molecular orbitals as opposed to atomic orbitals. For molecules with several atoms, the bonding tends to be delocalized over many of the atoms because orbitals on several of the atoms can overlap to form the molecular orbitals that then extend over several atoms in the molecule and not just the two involved in a localized bond.

The fact that several orbitals on several atoms can overlap to form new molecular orbitals derives from the fact that only atomic or molecular orbitals of the same symmetry can overlap, but all orbitals of the same symmetry in the molecule can overlap. Orbitals are composed of one or more lobes that have signs that are determined from the wave function of the given orbital. The combination of two orbitals to form a molecular orbital can only be bonding if the overlapping lobes have the same sign. If their signs are opposite then the net interaction will be antibonding. The overlapping lobes will have the same sign only if the point-group symmetries of the overlapping orbitals are the same (*42*).

A special case of molecular orbital formation is the situation in which two orbitals on one molecular fragment have the same symmetry; thus, both can overlap with a single orbital of the same symmetry on another fragment. This situation is opposed to the dihydrogen case in which only one level on each atom or fragment contributed to the bond. The resulting orbital picture is shown qualitatively in Figure 2.3 (*43*), where mixing the orbitals results in three new molecular orbitals. Of the three orbitals, one is bonding (Figure 2.3a), one is nonbonding (Figure 2.3b), and one is antibonding (Figure 2.3c). If within any of the three levels there are only two electrons, then the interaction will be net bonding because only the bonding orbital will be filled; with four electrons the nonbonding level will be filled, but the interaction will still be net bonding. With six electrons in the three orbitals the antibonding level will be filled; thus the interaction will be net antibonding and no bonding will take place. This special case will have application in Chapter 4 of this book.

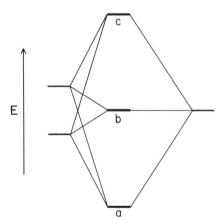

Figure 2.3. Molecular orbital picture for the overlap of three fragment orbitals of the same symmetry. Two lie on one molecular fragment. The three resulting orbitals are (a) bonding, (b) nonbonding, and (c) antibonding.

2.3   Descriptions of Metal Complex Bonding Theory

The basic picture of the bonding of the ligands to a transition metal atom is described in simple terms by crystal field theory and to a more realistic degree by ligand field theory (*44*). The two theories do not provide an exact description of the bonding in metal complexes; however, they do provide a general qualitative picture of the bonding that can explain conceptually, and with a fair degree of accuracy, the spectroscopic data that is seen in the study of transition metal complexes (*44*).

The crystal field theory postulates that the bonding in transition metal complexes occurs only through interactions of charge, that is, the bonding is described only by either ionic attractions or repulsions. Consider a transition metal ion, such as Co^{3+}, that requires 12 electrons to fill its 18-electron shell. The metal atom therefore requires six ligands, each donating two electrons, such as the common chlorine anion ligand (Cl^-), in order to fill its valence shell. For these six negatively charged ligands to both bond to the metal atom and repel each other as much as possible, they will arrange themselves about the metal atom so that each ligand lies at the apex of an octahedron (Figure 2.4). When this situation occurs, the five *d* orbitals are perturbed, and their degeneracies are lifted. The two *d* orbitals of the Co^{3+} ion that point to the apices of the octahedron ($d_{x^2-y^2}$ and d_{z^2}) are destabilized relative to unperturbed *d* orbitals; the remaining three *d* orbitals (d_{xy}, d_{xz}, and d_{yz}), which point away from the apices of the octahedron, are stabilized relative to unperturbed *d* orbitals (Figure 2.5). The resulting energy difference between the stabilized orbitals (t_{2g}) and the destabilized levels (e_g) is labeled Δ_0 and can in fact be measured very accurately in complexes by determining the energy required to promote an electron from the t_{2g} to the e_g levels. When electrons are placed in the *d* orbitals, the t_{2g} levels are usually filled first, and, as will be seen later, the t_{2g} levels provide the electrons for back bonding.

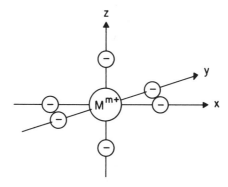

Figure 2.4. Octahedral arrangement of negatively charged ligands about a positively charged metal center.   (Reproduced with permission from reference 44. Copyright 1976 Wiley.)

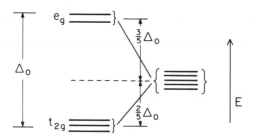

Figure 2.5. *d*-orbital energy splitting due to octahedral ligand arrangement. (Reproduced with permission from reference 44. Copyright 1976 Wiley.)

The value of Δ_0 is used as a measure of the bonding strengths of ligands because some ligands, by virtue of their coordination bond strengths, cause the t_{2g} and e_g orbitals to split to a large extent, whereas for other ligands, the orbitals are split only slightly. For example, the halides (I^-, Br^-, and Cl^-) lead to only small values of Δ_0; intermediate values are derived from ligands such as acetate (H_3CCOO^-), water, and glycine [$H_2N(CH_2COOH)$]; and larger values are obtained from ligands such as CO and cyanide (CN^-). In some cases, the energy splitting is so small that it is exceeded by the energy required to pair electrons in a single orbital. For those systems, after each t_{2g} level is filled with a single electron, the e_g levels are populated with one electron each before electron pairing occurs in the t_{2g} levels (*45*).

The more accurate ligand field theory is an extension of crystal field theory that allows for ligand–metal orbital overlaps (*44*). The six lone pairs of electrons, octahedrally displaced about the metal atom, require six empty orbitals on the metal to accept the electron density. In ligand field theory, these are the already mentioned $d_{x^2-y^2}$ and d_{z^2} orbitals and also the empty s, p_x, p_y and p_z orbitals, all rehybridized to form six orbitals oriented toward the apices of the ligand octahedron. The t_{2g} levels are unaffected and become available for overlap with unfilled ligand orbitals for ligand π-type back bonding (*44*).

Although the discussion so far has referred only to octahedral complexes, a similar treatment can be applied to complexes with lower symmetries and yields similar results. For other symmetries, however, the d orbital splitting will be different in both the relative ordering of the resulting perturbed d orbitals and the energy splittings between the levels.

The important point to be realized from this discussion, however, is that even though the ligands in organometallic compounds may all bond to a single metal atom, there are no large interactions between the ligands. The bonding of each ligand on the single metal atom is still similar to what is expected for a gas molecule on an on-top site on a surface, and if a ligand bonds to a monometal center, it should also bond to a surface in a generally similar fashion. Thus, the concept of a coordination site on a transition metal complex, that is, the place where a ligand is attached to the metal atom in a complex, can be considered analogous to the concept of surface site. Although the orbital structure versus band structure issue has not been addressed in this argument, it will be discussed in Chapter 5.

2.4   Direct Ligand-Ligand Interactions and Through-Metal Effects

Although no large interaction occurs between the ligands, they do interact with each other through the metal atom by second-order effects. Interactions similar to these may also occur on surfaces. One of these interactions is the inductive effect in which ligand species of high electronegativity can stabilize other orbitals on other ligands by the attraction of the ligand's orbital electron density. Electron spectroscopy for chemical analysis (ESCA) of a series of metal complexes in which the electronegativity of one of the ligands in the complex is changed can easily detect this stabilization effect (*36*) on other ligands in the complex. A similar effect occurring through a surface between two chemisorbed species would not be unreasonable to postulate.

Another effect that one ligand can have on another ligand in a monometal complex is called the *trans* influence (*45*). Although the effect is not well understood, it is apparent that a given ligand can influence the ground state properties, usually the metal–ligand bond length, of another ligand that is bonded *trans* to it in the complex. The word *trans* refers to ligands occupying opposite apices in the octahedron. As an example, X-ray crystallography has shown that the bond length of a Pt–Cl bond in a complex can be increased as the π back bonding ability of the ligand *trans* to the chloride ligand in the complex is increased. So, when the *trans* ligand is chloride, the Pt–Cl bond length is 2.317 Å; when ethylene, a better back bonder, is the *trans* ligand, the bond length increases to 2.327 Å; and when triethylphosphine $[P(C_2H_5)_3]$ is the *trans* ligand, a still better back bonder, the Pt–Cl bond length becomes 2.382 Å (*45*). The most likely explanation for this effect lies in the fact that for a monometal center, the amount of electron density that fills the back bonding d orbitals, the t_{2g} set, is limited, and as the π back bonding ability of the ligand in a given coordination site improves, it will deplete the back bonding in the coordination site *trans* to it and lead to the metal–ligand bond lengthening seen for the chloride ligand (*45*). This effect may also have a surface analogue because back bonding adsorbates must share a limited amount of electron density in the surface band structure. Adjacent ligands (*cis* ligands) exert a similar effect, but the magnitude is much smaller.

2.5   Ambidentate Ligands

One aspect of coordination chemistry that deserves special mention is ambidentate ligands (*46–48*) and their possible relationship to chemisorption experiments. *Ambidentate ligands* are defined as molecules that can bind to a transition metal coordination site through more than one atom in the molecule. Ambidentate ligands are quite common in coordination chemistry, and it seems likely that these moieties, under the proper conditions, could exhibit a similar behavior on surfaces. Examples of this type of ligand will appear throughout the remainder of this book. These ligands can be exemplified by a molecule such as pyridine, which can bond through either the lone pair of the nitrogen atom or the aromatic *p* orbitals (*49*). Other ambidentate ligands that may be less familiar include the nitrogen dioxide ligand, which can bond through either the nitrogen atom (Structure **2.2a**) or through one or both of the oxygen atoms (Structure **2.2b**) (*46*), and the thiocyanate anion (SCN$^-$), which can bond to a metal center through either the nitrogen atom (Structure **2.2c**) or the sulfur atom (Structure **2.2d**) (*48*). These ligands may be of interest with regard to chemisorption studies because it is likely that a great deal about the nature of the chemisorption bond can be learned by studying the bonding characteristics and preferences of ambidentate ligands on surfaces.

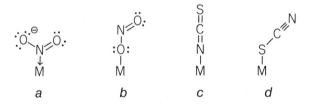

Structures **2.2a–2.2d**. Examples of ambidentate coordination for (a and b) the nitrite (NO$_2^-$) ligand and (c and d) the thiocyanate (SCN$^-$) ligand.

To account for this ambidentate behavior, several theories have been proposed. The simplest is a steric argument; that is, a ligand will bond through a slightly less stable bonding interaction if the thermodynamic destabilization of the complex caused by crowding of the ligands about the metal atom can be reduced (*47*). The nitrogen dioxide ligand will usually bond through the nitrogen atom; however, in the presence of large bulky neighbor ligands, NO$_2$ may bond through the oxygen atom to form the nitrito ligand, which requires less coordination volume around the metal center (*47*). For thiocyanate (SCN$^-$), the N-bound species is linear and requires less volume than the S-bound species, which is usually bent and thus requires a greater coordination volume (*48*).

Electronic effects (*47, 48*) have also been proposed to account for the various bonding modes of ambidentate ligands. One of those effects is the relative π back bonding of each atom in the ligand. For thiocyanate (SCN^-), the nitrogen end is a poor π electron acceptor and thus is unable to accept π electron density from the metal. The sulfur atom, on the other hand, is a relatively good π electron acceptor. In complexes in which a high electron density is found on the metal, an S-bound SCN ligand is preferred because it can stabilize some of the excess negative charge on the metal through π back bonding. When the electron density on the metal is low, the N-bound moiety (the poorer π back bonding species) is expected to form. This hypothesis is supported by the two palladium dithiocyanate complexes shown as Structures **2.3a** and **2.3b**. In Structure **2.3a**, the amine ligands (NH_3) are only σ donors; because they augment the electron density on the Pd atom and thus its π donor ability, the S-bound thiocyanate ligand (the good π back bonder) is preferred. In Structure **2.3b**, the triethylphosphine ligands (PEt_3) are very good π acceptors. They deplete the Pd atom of electron density, and thus the N-bound thiocyanate ligand (the poorer π acceptor) is obtained (*48*).

$$\begin{array}{ccc} H_3N & & SCN \\ & Pd & \\ H_3N & & SCN \end{array} \qquad \begin{array}{ccc} R_3P & & NCS \\ & Pd & \\ R_3P & & NCS \end{array}$$

a   *b*

Structures **2.3a** and **2.3b**. Illustration of ligand dependence on the coordination of ambidentate ligands: (a) NH_3 ligands soften the metal center, favoring interaction with the soft end of the SCN^- ligand; (b) the π acceptor ligand PR_3 hardens the metal center, favoring interaction with the hard end of the SCN^- ligand.

A second electronic effect that may affect the bonding in ambidentate ligands involves what is called the hard–soft acid–base (HSAB) theory (*46*). Orbital overlaps can be maximized when the two given orbitals are close in energy. HSAB theory states that as a second order effect, orbital overlaps can be further improved if the overlapping orbitals are generally of the same size. Atoms with small orbitals are considered to be hard, whereas atoms with large orbitals are considered to be soft. Hard atoms with small orbitals tend to bond with other hard atoms, whereas soft atoms with relatively large orbitals tend to bond with other soft atoms. The ambidentate bonding in the thiocyanate ligand can be explained by HSAB theory, which identifies nitrogen as a hard atom and sulfur as a soft atom. The electron-donor ligands (NH_3) seen in Structure **2.3a** soften the Pd atom by increasing its electron density and thus its orbital size. These changes increase its propensity to bond with the soft S atom. The electron-withdrawing ligands (PEt_3) seen in Structure **2.3b**, harden the Pd atom by shrinking the metal orbitals and increase its propensity to bond with the hard nitrogen atom (*46*).

Ambidentate ligands are especially interesting because they can form linkage isomers, that is, transition metal complexes that have the same set of ligands but differ in the mode of attachment of one of the ambidentate ligands (46). In these cases, one isomeric form is the thermodynamically stable species, whereas the second is a kinetically stable species (47, 48), a species that requires a large activation energy to isomerize into the more stable thermodynamic form. Conversion back into the kinetically stable species can sometimes be induced photochemically (46). This kind of isomerism probably could be carried out on surfaces under the proper conditions.

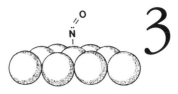

3

Overview of Coordination Ligands

3.1 Introduction

THE THIRD PART OF THIS BOOK IS A DISCUSSION of the various molecular species commonly used as ligands in coordination chemistry. This discussion is intended as a means of categorizing the variety of species that could potentially be studied on surfaces. Although many have already been examined in a cursory way on surfaces, none have been studied in the depth accorded the CO ligand. This situation is an indication of the potential for research that each of these moieties can provide. Again, because these species bond to transition metal atoms in complexes, they should also bond in a qualitatively similar fashion to transition metal surfaces. This behavior has already been confirmed experimentally for a few of these species such as CO (*8, 9*) and acetylene (*11, 15*).

Although the molecular structures of the various ligands may differ when they are bound to a metal atom, the fundamental nature of metal–ligand bonding itself seems to be rather general from ligand class to ligand class. Specifically, there appear to be relatively few examples of metal–ligand bonding involving only a single bond. For the most part, the double-bond character found in CO metal complex bonding and chemisorption is generally the rule, that is, σ bonding by lone-pair donation to the metal and concomitant π back donation from the metal into unoccupied π states on the ligand (Figure 3.1). Deviations in this type of bonding are found, however, because the σ- and π-type donations can also occur in a manner reverse to that of CO. Ligand-to-metal π donation is known in several ligand classes such as the alkylamides (*50*) and metal-to-ligand σ donation is thought to be present in species such as bent NO and diazo (N=N–R) ligands (*51*).

1003–9/87/0019$22.20/1 © 1987 American Chemical Society

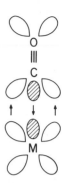

Figure 3.1. Fundamental picture of CO coordination to a metal center. Shaded lobes illustrate $5\sigma \rightarrow d$ metal donation, whereas unshaded lobes illustrate $d \rightarrow \pi^*$ back donation.

In addition to bonding generalities in complexes, general trends in the bonding of gases on surfaces can be seen. Although data concerning the chemisorption of many ligand species on surfaces is quite scarce, these interactions tend to fall into three general categories. First, the bonding can be essentially the same on a surface and in a complex. Examples of this type of bonding include terminal CO bonding (52) and the π-type bonding of acetylene (53). A second class of surface bonding involves an interaction that can be placed along a continuum of interactions lying between two extremes. The bonding of ethylene (54) is an example of this class where rehybridization of the carbon atoms could be either significant, leading to substantial out-of-plane bending of the C–H bonds, or nonexistent, resulting in a planar ethylene ligand, depending upon the nature of the surface. Bonding in a Dewar–Chatt–Duncanson sense (55, 56), that is, ethylene π-to-metal d donation and d-to-π^* back donation, is always present; the only differences in bonding are due to the relative magnitudes of the two bonding interactions. The third class involves ligands whose bonding is generally different between surfaces and complexes. When this type of bonding arises, however, the effect can most often be attributed to the presence of the vacant surface sites that are adjacent to the metal atom involved in the surface bonding. Examples of this kind of interaction include atomic oxygen and atomic nitrogen ligands, which bond exclusively to monometal centers in coordination complexes (57, 58) and to hollow sites on surfaces (59, 60). This condition may be a thermodynamic effect in which these electron-donor ligands apparently tend to maximize their metal atom coordination on surfaces to minimize the energy of the system.

3.2   General Methods of Preparing Surface-Chemisorbed Species

An important aspect of this work involves not only describing the bonding of ligands on surfaces but also delineating the methods that can be used to place the ligands on the surface. These methods will be discussed briefly before the individual ligands are discussed. These surface preparative techniques can be generalized into five groups that are generally unrelated to coordination chemistry syntheses because of the great differences in the two types of systems. The preparative methods used in organometallic and coordination chemistry syntheses generally involve solution-phase reactions and ligand substitutions, which are not analogous to chemisorption from the gas phase onto clean surfaces in ultrahigh vacuum. More importantly, many of these organometallic synthetic techniques leave behind other impurities that could be easily removed in purification procedures in chemical synthesis but would often remain on a surface and lead to a nonhomogeneous overlayer. This listing does not include preparative techniques used in chemisorbing species to electrodes in the solution phase. These methods are much more straightforward and have not yet been applied to any large extent (*61*) to single-crystal systems in an ultrahigh-vacuum (UHV) environment.

The most common method of surface preparation is the exposure of a clean surface to a gas. This technique is used for the chemisorption of CO, NO, ethylene, etc. This method, however, has an even wider applicability because it is not limited to substances that are gases at room temperature. Because exposures to clean single-crystal surfaces are rarely done at pressures above 10^{-6} torr, any liquid or solid with a vapor pressure exceeding 10^{-6} torr can be introduced to a clean surface by this volatile gas method. Such liquids and solids encompass an enormous range of organics and some inorganic compounds as well. Ligands such as benzene (*19, 62, 63*) and acetic acid (*64*) are two examples of this kind of species. Highly volatile impurities in the adsorbate stock can be a hindrance during dosing because they will preferentially transfer in an exposure. This problem can be overcome, however, by using highly pure samples and careful handling.

The second type of surface preparation method is best referred to as hydrogen capping. Hydrogen capping is not a new concept; it does, however, consolidate several ideas into a single surface synthetic procedure. Hydrogen capping refers to a method of introducing to a surface certain ionic species that are not volatile but that can be made volatile by using a hydrogen ion as the counter ion. For example, the cyanide anion, a very common coordination chemistry ligand, is a nonvolatile species when bound to sodium in NaCN. Although coordination chemists can easily prepare complexes in solution by using the highly soluble NaCN moiety, it is useless as a means of preparing a surface overlayer because the nonvolatile anion cannot be placed on a clean surface in a vacuum. However, if the cyanide anion is *hydrogen capped,* that is, bonded to hydrogen instead of sodium, the anion then becomes gaseous and thus can be transported to a clean surface in a UHV environment. Once on the surface, the H–C bond may be broken through the proximal effect as described by Muetterties (*65*).

a
$$\overset{H}{\underset{H}{\nwarrow}}\overset{}{\underset{\nearrow}{C}}-O\overset{H}{\nearrow} \qquad \xrightarrow{\text{SURFACE}} \qquad \underset{|}{\overset{O}{\underset{C}{\|}}}$$

b
$$\underset{H_2N\!\cdot\quad\cdot NH_2}{\overset{H_2C-CH_2}{\diagup\quad\diagdown}} \qquad \xrightarrow{\text{SURFACE}} \qquad N{\equiv}C-C{\equiv}N$$

Schemes 3.1a and 3.1b. Two examples of the hydrogen-stripping adsorbate preparation method reported in the literature: (a) preparation of surface CO from methanol and (b) preparation of surface cyanogen from ethylenediamine.

This breakage would leave cyanide fragments on the surface for characterization and hydrogen atoms, which are usually weakly bound to the surface and could probably be displaced by other cyanide ligands. Other examples of the use of hydrogen-capped adsorbates include methoxy (66), which is prepared from its hydrogen-capped counterpart methanol, and surface acetate, which is prepared by exposing a surface to acetic acid (67), the hydrogen-capped acetate ligand.

A third surface preparation method might be called hydrogen stripping. In this method, the ligand to be chemisorbed on the surface is made the core of a molecule that is surrounded by hydrogen atoms. When this molecule is placed on the surface, all of the hydrogens are stripped away from the molecule; this stripping leaves the core of the molecule, which becomes the surface chemisorbed species. A trivial example of this process might be the preparation of surface CO through the decomposition of methanol (Scheme 3.1a) (68). The core of the methanol molecule is C–O, and it is surrounded by hydrogen, which gets stripped away upon chemisorption. A similar effect is seen for ethylenediamine (H_2N–CH_2–CH_2–NH_2) (Scheme 3.1b) (69a), which decomposes or undergoes hydrogen stripping to form cyanogen (N≡C–C≡N). Partial hydrogen

Structures 3.1a and 3.1b. Structure of (a) benzylidyne prepared by partial hydrogen stripping during the decomposition of (b) surface toluene.

stripping would also seem possible, as is proposed in the formation of benzylidyne (Structure **3.1a**), which may be prepared from toluene vapor (Structure **3.1b**) (*65, 69b*) through partial hydrogen stripping of the methyl hydrogens only. These examples are only trivial cases, but it may be possible, through this technique, to prepare on surfaces new species that can be prepared by no other method.

The remaining two general surface preparative methods include the use of dimeric species and surface reactions. The use of dimers is similar to hydrogen capping in that nonvolatile species are made volatile, not through hydrogen capping but through dimerization. Continuing with the cyanide analogy, cyanide dimer, that is cyanogen ($N\equiv C-C\equiv N$) is also gaseous and can be used as a method of introducing CN moieties onto a clean surface (*70, 71*). The only requisite is that the dimer undergo scission on the surface to its monomers, an effect that can only be determined experimentally but is known to occur for cyanogen (*71*).

The final method is by far the least common surface preparation method in that a surface reaction of one or more molecules produces the desired ligand. Hydrogen capping, hydrogen stripping, and the use of dimers are subsets of this more general technique, but the technique encompasses a much broader range of reactions. Examples of this method of preparation include surface methylene ($=CH_2$) which has been prepared on Ru(001) through the chemisorption and reaction of diazomethane ($N=N=CH_2$) (*72*), and surface SO_3, NO_2 and NO_3 moieties, which have all been prepared on powdered silver by reacting surface oxygen with gaseous SO_2 (*73*), NO (*74*), or NO_2 (*74*) species. Of the four surface preparative techniques discussed, the surface reaction method probably holds the greatest potential in preparing new surface species.

3.3   Ligand Classes

The various ligands that will be discussed in the rest of this chapter fall under the following 13 classes:

1. Lone-pair donor species (ammonia, water)
2. Diatomics (CO, NO)
3. Substituted diatomics (methyl isocyanide)
4. Cumulenes (allene, CO_2)
5. Heteroolefins ($H_2C=O$, $R_2S=O$, etc.)
6. Triatomic pseudohalogens ($N=C=O^-$, nitrous oxide)
7. Chelates (acetate, ethylene diamine)
8. Nonmetal oxyanions (NO_2^-, $SO_3^=$)
9. Aromatics (benzene)
10. Heterocyclics (pyridine)
11. Electron-deficient ligands (hydride, boron hydrides)
12. Molecular-fragment ligands (ethylidene)
13. Atomic species (Cl^-, S_2^-).

Although other classifications may be possible, the bonding modes of the ligands in each class are similar. Grouping the ligands according to the atom in the ligand that coordinates to the transition metal is avoided in this book because the structures and bonding interactions of coordinated ligands are more important to surface science research than the coordination chemistry of a given element. For this same reason, every ligand discussed will not necessarily fit into the name given its class. The class names apply only to the majority of the species in the class. The bonding characteristics of each ligand in a given class, however, will be similar.

In a review of this nature, it would be impossible to cover all the primary papers for the various ligands because that would encompass almost all of the coordination chemistry literature. Instead, this chapter has been limited to a discussion of the review literature that discusses in more general terms the bonding and methods of characterizing the various ligands. The primary literature has, however, been reviewed where appropriate. In addition, the surface science literature of a given ligand has also been reviewed within the discussion of that ligand as a means of presenting exemplary surface investigations of the molecular species being discussed. Although every appropriate surface science investigation is not represented, it is believed that every species that has been examined on a surface is at least mentioned. Although few direct comparisons are made between the organometallic literature and the surface science literature for a given ligand, the similarities of structure and bonding appear to be quite compelling.

One very important issue that cannot be covered in detail in this chapter concerns the thermodynamic differences encountered between monometal complexes and multiatom surfaces. The bond energies for some of the ligands to be discussed are so small that it is likely they will dissociate to adsorbed atoms on a chemically aggressive surface. Organometallic chemistry provides no insight into surface dissociation effects, and any attempt to draw some dividing line between bond energies that might lead to dissociative or associative adsorption on a particular metal surface would be specious. However, any dissociative process on a surface requires a low-energy pathway to occur. Indeed, because a broad range of surface reactivities is known for the transition metals, most of the species will not likely dissociate on the least active surfaces, such as those of copper, silver, and gold.

3.4   Lone-Pair Donor Ligands

The simplest class of ligands are those molecular species that donate a nonbonding lone pair to the transition metal. If there is no back bonding, these species generally bond only weakly to the metal center and usually require the metal to be in a high-valence state for bonding to occur. They usually bond in a terminal fashion. Ligands in this group include species such as water (Structure **3.2a**) (*75*), alcohols (Structure **3.2b**) (*76*), ethers (Structure **3.2c**) (*77*) and ammonia (Structure **3.2g**) (*77*). Such ligands undergo no back bonding because the σ-bonded oxygen and nitrogen donor atoms have no low-lying acceptor orbital, such as the π level in CO or acetylene. These species, when bound in a complex, show only minor changes in their infrared (IR) spectra relative to the free species. The changes normally involve a slight decrease in the symmetric and asymmetric stretching frequencies of the bonds adjacent to the donor atoms and an increase in the deformation rocking-mode frequencies (*76, 77*). These changes are inversely related to the strength of the metal–ligand bond. Most ligands of this type are either gases or volatile liquids and can easily be adsorbed on a clean surface.

Structures **3.2a-3.2g**. Ligands that coordinate through nonbonding or lone electron pairs.

For similar species in which sulfur is the donor atom (*78*), the bonding is stronger because of a small back bonding ability that results from the empty low-lying *d* orbitals on the sulfur atom. Ligands of this type include hydrogen sulfide, thioalcohols (mercaptans), and thioethers (Structures **3.2d-3.2f**). The metal–sulfur stretching frequency is found between 200 and 500 cm^{-1} (*77*).

When back bonding occurs in addition to lone-pair donation, the coordination of nonbonding lone-pair donors becomes strong even though σ donation remains the predominant bonding effect (*79*). In the case of the phosphines (Structure **3.3a**) (*79*) where R can be almost any kind of organic group, the back bonding brings the metal–ligand bond strength up to a magnitude near that of CO, and it becomes possible for phosphines to stabilize low- and zero-valent metal oxidation states as well as the higher oxidation states stabilized by other lone-pair donors (*79*). Various substituted phosphines such as triethylphosphine (Structure **3.3b**) and, most commonly, triphenylphosphine (Structure **3.3c**) are used almost as widely as CO in organometallic complexes. The changes in the vibrational spectra of phosphine ligands that occur with coordination are similar to the changes for NH_3 in that the P–R stretching frequency changes only slightly upon coordination. The symmetric P–R stretching frequency, however, is observed to increase rather than to decrease (77, 80). Phosphine itself is a gas, whereas triethylphosphine and triphenylphosphine are a volatile liquid and a low-melting solid, respectively (81). Therefore, these species can quite easily be placed on a clean surface by simple vaporization.

Structures 3.3a-3.3e. Bonding modes of phosphine ligands and several exemplary phosphines; ϕ refers to a benzene ring.

Other phosphine molecules used as ligands are the phosphites (Structure 3.2d) and the dimethylaminophosphines (Structure 3.2e) (79). In addition, PF_3 is used as a ligand because its highly electron-withdrawing fluorine substituents stabilize the empty d orbitals of the phosphorus atom and thus lead to a substantially enhanced back bonding and a stronger metal–phosphorus bond (82). The asymmetric and symmetric stretching frequencies of PF_3 fall near 850 and 750 cm^{-1}, respectively, when bound in a complex (77).

Several lone-pair donor-type species have been chemisorbed on single-crystal surfaces. NH_3 has been studied on Ni(110); a Ni–N stretch has been detected at 570 cm^{-1} with electron energy loss spectroscopy (EELS) (83), and angle-resolved ultraviolet photoelectron spectroscopy (ARUPS) work (84) has determined that the Ni–N bond is normal to the surface. Electron-stimulated desorption ion angular distribution (ESDIAD) work (85) indicated that the ammonia molecule occupies an electropositive ridge atom site. On Ir(111), ARUPS work has determined that NH_3 lies in the three-fold hollow with a vacancy underneath (*86*). A similar molecular adsorption of substituted amines has also been reported (*87*).

H_2O has been studied on Pt(111) (*88*), Ti(001) (*89*), oxygen-covered Ru(001) (*90*), Cu(100) (*91*) and Pd(100) (*91*) and other metals. On Pt and Ti, O–H dissociation was detected; on the oxygen-covered Ru(001), an O–H_2O complex was formed. For the Cu(100) and Pd(100) surfaces, EELS has shown that the molecular axis of the molecule is tilted relative to the surface normal (*91*). Three calculations concerning chemisorbed water have also been reported (*93–95*); these found that H_2O interacts with the surface through its lone pairs (*94, 95*) and that as the interaction increases, the HOH angle in the molecule increases (*94*). The HOH angle is believed to increase because of image forces (*93*).

Alcohols adsorb without dissociation on polycrystalline palladium (*96*), Fe(110) (*97*), and Cu(111) (*97b*). On Rh(111) (*98*) and Ru(001) (*99a*), methanol dissociates to only a small extent at 100 K. The overlayer showed no long-range order, but two chemisorbed states were detected by temperature-programmed desorption (TPD). Decomposition to CO was seen by an electron loss spectroscopy (ELS) loss for CO at 13.7 eV. On both clean and oxygen-modified Mo(100) (*99b*), methanol also decomposed to CO and hydrogen; however, as the coverage of oxygen increased on the surface, the decomposition reaction diminished. Surface methoxy species were detected on sodium precovered Cu(111) (*97b*). In a recent work (*100a*), TPD was used to study the adsorption of a variety of weakly adsorbing alcohols and ethers on Cu(100) and Pt(111). There is one report of the chemisorption of methanethiol (H_3C–SH) on Pt(111) (*100b*). High-resolution electron energy loss spectroscopy (HREELS) and near-edge X-ray absorption fine structure (NEXAFS) were used to characterize both thiomethoxy (H_3C–S–) and thioformaldehyde adsorbates as intermediates in the decomposition of methanethiol on Pt(111).

Several studies of phosphine (PR_3) chemisorption have been reported. Trimethylphosphine has been adsorbed on Ni(111) as a method of displacing preadsorbed benzene (*101*). Trifluorophosphine (PF_3) (*102*) has been examined on a variety of surfaces by using ultraviolet photoelectron spectroscopy (UPS), ELS, low-energy electron diffraction (LEED) and TDS. It was found that PF_3 bonds to a surface through the phosphorus atom in a manner analogous to coordination compounds and that the chemisorption bond is localized. Phosphine (PH_3) itself has been studied on Rh(100) by using a similar array of techniques (*103a*). Work function measurements showed a 1.2-eV decrease in the work function. This result indicated adsorption through the phosphorus lone pair whereas TPD detected the evolution of H_2 and PH_3; the coverage dependence of the adsorption energy indicated repulsive lateral interactions of chemisorbed PH_3 moieties. Coadsorbed species such as hydrogen, deuterium, and oxygen seemed to have no effect on the PH_3 ligand (*103b*).

3.5   Diatomics

The diatomics are probably the most widely studied and well-understood ligands in coordination chemistry. These include CN^-, CO, N_2, NO, and O_2 as well as C_2H_2, which, although not rigorously a diatomic, bonds in ways generally similar to diatomic species. What is most interesting about diatomic ligands in coordination chemistry is that they form a series between linear-bonded, bent-bonded, and flat-lying species. This behavior may provide insight into the nature of the bending interaction of diatomics on surfaces. The first part of this section will be a discussion of the individual diatomic ligands and their bonding modes; the second part will discuss molecular orbital calculations that relate the various bonding modes of diatomic ligands to better understand the effects of diatomic bending and postulate how this behavior may apply to surfaces. Each diatomic ligand is presented in this section in order of increasing propensity for bending (*104*), an order that will be significant in the second part of this section.

The diatomics can also be grouped together because their molecular orbital structures are all qualitatively the same as that of CO, the only difference being the relative energies of the individual orbitals. The differences in the bonding of these diatomics is due essentially to these energy differences, which can make a given bonding mode more stable than another for a given set of energy levels.

The bonding for all of the diatomic species is much like the well-known situation for CO (Figure 3.1) (*45*). The 5σ orbital of the CO ligand donates electron density to the metal, whereas the filled d_{xz} or d_{yz} levels on the metal back donate electron density into the 2π level (the π^* orbitals) of the CO molecule. The bonding can be characterized with vibrational spectroscopy because the back donation weakens the CO bond and results in a lowering of the CO stretching frequency. For the other diatomics, the differences in their bonding relative to CO can be traced to the 5σ and 2π orbital energy levels; if the 5σ orbital is higher in energy than that of CO, the diatomic becomes a better σ donor; if the 2π orbital on the diatomic is lower in energy relative to the 2π on CO, it becomes a better π acceptor. The reverse of these arguments also holds. For flat-lying species, the only difference is that the orbital energy differences make the π-bond orbital of the diatomic (the 1π level) the better σ donor relative to the 5σ level. Diatomic bending can be rationalized from a similar kind of argument that will be discussed next.

3.5.1   Cyanide

The cyanide anion is known to bond to transition metals in the terminal (Structure **3.4a**) and bridging (Structure **3.4b**) modes (*105*) as well as a linear bridging mode (Structure **3.4c**) (*105, 106*) in which both ends of the ligand are bound to metal atoms. The M–C≡N bond angle can be as low as 160°. The isocyanide linkage, with bonding solely through the nitrogen atom, is unknown because the higher electronegativity of the nitrogen atom stabilizes its lone pair to σ donation (*105*). Bonding by N coordination is known, however, for the neutral HCN ligand (Structure **3.4d**) (*107*), which, when deprotonated, isomerizes rapidly to the carbon-bonded species (*107*).

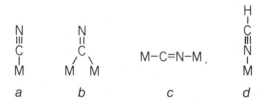

Structures **3.4a-3.4d.** Bonding modes of the cyanide anion and hydrogen cyanide gas.

Unlike CO, coordinated CN⁻ can show wide differences in its C≡N stretching frequency because its bonding can be either predominantly σ in nature, with little or no π back bonding, or with substantial back bonding in addition to the σ bonding (*105*). These differences from CO bonding are attributable to the fact that both the 5σ and the 2π orbitals on CN⁻ lie to much higher energy than the corresponding orbitals of CO (*104*). When σ bonding predominates, the C≡N stretch increases by up to 100 cm⁻¹ relative to its gas-phase value of 2078 cm⁻¹; this result occurs because the 5σ level, the lone-pair donor on the CN– ligand, is slightly antibonding with respect to the C≡N bond. As electron density in the 5σ level is reduced through coordination to the metal, the bond order of the C≡N bond increases, and thus the vibrational frequency increases. When substantial π back bonding occurs, the C≡N bond order and thus the C≡N stretching frequency are reduced in a manner identical to that for CO (*105*). For linear bridging CN– ligands, as seen in Structure **3.4c**, the C≡N stretch is usually 50 cm⁻¹ higher in frequency than the terminal ligand. This phenomenon is very useful in detecting the presence of this kind of bridge (*105*). The N-coordinated HCN ligand is characterized by the presence of the C–H bond stretch near 3200 cm⁻¹, which is lower in frequency relative to the gas-phase value. The C≡N stretch frequency will usually increase for this bonding mode (*107*). The extensive amount of IR data characterizing the CN– ligand is complemented by electron spectroscopy for chemical analysis (ESCA) studies (*108*).

The cyanide anion has been studied on Ru(001) (*70*), Pt(111) (*65*), Cu(111) (*109*), Ag(111) (*110*) and Rh(111) (*111*). In each study, cyanogen was used to deliver the CN ligand to the surface, and it dissociated at least in part on each surface. In addition, undissociated cyanogen was detected on each of the surfaces at least at low temperature. On Rh(331), surface cyanide was prepared by the reaction of nitric oxide with preadsorbed carbon at temperatures greater than 400 K (*112*). There are also a few reports of HCN and cyanogen chemisorption on supported dispersed metals (*107*). For both dispersed nickel and iridium, the CN stretch bands are all at a frequency higher than that of gas-phase hydrogen cyanide; this finding indicates the predominance of σ donation in the metal–ligand bond. The metal–cyanide bond has also been studied theoretically (*113*); the predominant interaction was shown to be a 5σ-to-metal s interaction with little or no d-orbital involvement. In addition, the derivative of the dipole moment was lower for coordinated cyanide than free cyanide and led to an expected attenuation in the IR extinction coefficient for bound cyanide over free cyanide (*113*).

From this discussion, it appears that hydrogen cyanide, HCN, would be a very interesting species to study in detail on surfaces. Through hydrogen capping, a CN^- ligand may be delivered to the surface. On the other hand, if no C–H scission took place on the surface, N-bound HCN may be detected, that is, the isocyanide structure. In this case it may be possible to deprotonate the species and thereby cause isomerization to the C-bound form. It may even be possible to coadsorb hydrogen with chemisorbed CN and generate the surface-bound HCN ligand.

3.5.2   Dinitrogen

For the case of dinitrogen (N_2), little, relative to CO, is known about its bonding in metal complexes, although N_2 has been studied in some detail. It is bound most commonly in transition metal complexes in a linear fashion to either one (Structure **3.5a**) or two (Structure **3.5b**) metal atoms (*114*). It is less commonly found side-on bonded (Structure **3.5c**) like ethylene or in the combination of terminal bonding and π bonding shown in Structure **3.5d** (*114*). Because the electron-pair donor orbital on dinitrogen, the $3\sigma_g$ has a lower energy than the 5σ electron-pair donor on CO, N_2 is a poorer σ donor than CO. Conversely, since the $1\pi_g$ orbital on N_2 is higher in energy than the π electron-accepting 2π orbital on CO, N_2 is a poorer π acceptor than CO (*115*). Both of these factors explain why N_2 bonds much more weakly than CO to transition metals in complexes. N_2 is further disfavored

$$
\begin{array}{cccc}
\begin{array}{c} N \\ \vert\vert\vert \\ N \\ \vert \\ M \end{array} &
M{-}N{\equiv}N{-}M &
\begin{array}{c} N{=}N \\ \diagdown\;\diagup \\ M \end{array} &
\begin{array}{c} \qquad N \\ \diagup\!\!\diagup \\ N \\ \diagup\quad\diagdown \\ M \qquad M \end{array} \\
a & b & c & d
\end{array}
$$

Structures **3.5a-3.5d**. Bonding modes of dinitrogen.

toward bonding because the lobes of its orbitals are symmetric with respect to the plane bisecting the N–N bond and will be unable to overlap with the metal levels as well as CO, whose orbitals are skewed in size toward the carbon end of the molecule (*115*).

An interesting side effect of this orbital symmetry is that the end-on bonded dinitrogen ligand can undergo an end-over-end rotation because it can coordinate in a terminal fashion through either nitrogen atom. The intermediate is apparently the side-bound mode. The process is not dissociative because the rate of substitution reactions of the N_2 ligand, in which the ligand must separate from the metal complex, is 45 times slower (*116*) than the rate of this end-over-end rotation. This finding means that when the linear dinitrogen ligand changes the atom that is bound to the metal center, the process occurs too rapidly to be caused by a ligand dissociation process.

Although both the σ and the π bonding in the dinitrogen ligand are weak relative to CO, the π bonding predominates. This finding is supported by the vibrational data of end-on bonded N_2, which show that the $N\equiv N$ stretch is 100–400 cm^{-1} lower than that for gas-phase dinitrogen (2331 cm^{-1}) (*115, 117*). Similar to the CN^- ligand discussed earlier, if σ bonding predominates, the $N\equiv N$ stretch will increase because the electron density will then be removed from the slightly antibonding $3\sigma_g$ orbital. Apparently, because of the *trans* effect discussed earlier, the $N\equiv N$ stretch can increase when other ligands in the complex are better π acceptors and deplete the metal atom of electron density (*115*).

In a recent theoretical study (*114*), it was proposed that N_2 could bond in much the same way as its isoelectronic analogue, acetylene, in the acetylene hexacarbonyl dicobalt complex (Structure **3.6**). The structure was thought to be reasonable for the dinitrogen ligand because the molecular orbital diagrams of both species are the same. In addition, the total bonding interaction in the N_2 species would be similar to the acetylene species because the back bonding that was lost to the dinitrogen ligand through poorer overlap relative to acetylene could be made up through a better orbital energy match between the orbitals involved in the back bonding (*114*). A compound with this mode of dinitrogen bonding, however, has not yet been prepared.

The adsorption of N_2 on surfaces is in some cases analogous to the bonding of the N_2 ligand in metal complexes. On Ni(110) (*118*) and Ni(100) (*119*), ARUPS was used to determine that N_2 chemisorbs to a single

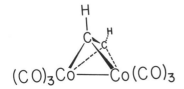

Structure **3.6**. Structure of the acetylene hexacarbonyl dicobalt complex showing the acetylene ligand bound in the π-bonded bridging fashion to two metal centers.

surface atom with its axis normal to the surface. This bonding mode was corroborated for dinitrogen on Ni(100) by X-ray photoelectron spectroscopy (XPS) from which it was found that the surface species had two N $1s$ binding energies and thus two inequivalent nitrogen atoms (*120*). On Ru(001), an EELS study found a similar mode of bonding for chemisorbed N_2 (*121*). On Pd(111), however, an ARUPS study found that the N_2 surface bonding was random (*118*). On Cr(110), dinitrogen was found to dissociate at 300 K and form a nitride overlayer (*122a*). On Fe(111), UPS and HREELS were used to characterize a flat-lying, π-bonded, dinitrogen moiety, a species that is thought to be a precursor to dissociation (*122b*). When nitrogen was adsorbed on potassium covered Fe(111), the heat of adsorption increased for bound N_2 near the potassium atoms (*123*). This result was attributed to an enhanced back bonding in the N_2 molecule due to the increased surface electron density caused by the electron-donating potassium atom. On dispersed rhodium, transmission IR found the N_2 stretching frequency to be shifted to lower frequency by 74 cm^{-1} relative to gaseous dinitrogen; this result indicated a weak interaction with the metal, much like that found in transition metal complexes. The surface bonding mode was not ascertainable (*117*).

3.5.3 Carbon Monoxide

The most commonly known diatomic is CO, carbon monoxide. Its common bonding modes include the terminal, two-fold bridging, and three-fold bridging modes (Structures **3.7a-3.7c**). Because CO is so commonly used and understood in the surface science community (*52*), there is little need to reiterate the basics of CO coordination. However, some relatively new developments in CO coordination chemistry merit the attention of surface scientists and will be discussed.

Unsymmetrically bridged CO species (Structure **3.7f**) are a fairly recent discovery in coordination chemistry (*124*). This type of bridging involves pairs of CO ligands and, as determined by X-ray crystallography, shows the CO bridging ligand with one metal–carbon bond length substantially longer than the other. The second CO ligand in the pair then shows a similar bridge bonding but is reversed with respect to the metal atoms. This system can be thought of as a frozen action picture of CO ligands moving from terminal positions on opposite metal atoms toward symmetrical bridging positions between the metal atoms. In these unsymmetrical bridges, the net electron density on each metal atom is the same as for symmetric bridge bonding because the average M–C bond lengths equal the bond length of a stable symmetric bridge. It is not impossible to conceive of this type of bonding on surfaces on which steric crowding could force adsorbed CO into asymmetric bridge sites.

One of the most interesting of the new CO bonding modes that have

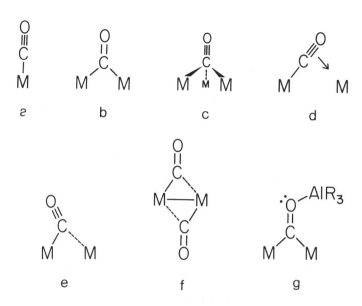

Structures **3.7a-3.7g**. Bonding modes of carbon monoxide.

been discovered is the semibridging CO ligand (Structure **3.7e**) (*124*). This species is distinguished from unsymmetrically bridged species by the fact that semibridging CO ligands do not come in pairs and have an M-C-O angle that is less than 180°. They are believed to form as a means of equalizing the electronic charge in unusual dimetal species. As mentioned earlier, all metal atoms in complexes try to achieve an 18-electron closed-shell configuration. For the cases in which a coordination complex requires only one additional electron to close its shell, it will tend to dimerize with another similar complex and form a metal–metal bond, gaining its one additional electron in the covalent metal–metal bond. If, however, one coordinated metal center requires two additional electrons, it could be forced to accept an electron pair from an adjacent metal atom in the complex. An unstable situation arises because the donor metal atom will have a formal charge that is higher than the acceptor metal atom. This separation of charge is alleviated by the acceptor metal atom donating electron density back to the donor metal atom via a π^* orbital of a CO ligand. This situation leads to the semibridged bond (Figure 3.2) (*124*). This type of structure is something that may be found on alloy surfaces on which surface metal atoms with different numbers of electrons may cause unusual metal-metal donor properties.

Besides the various modes of carbon-bonded CO, there are also two types of coordinated CO that bond not only through the carbon atom but also through the oxygen atom (*125*). This situation involves two types of

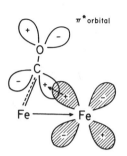

Figure 3.2 Illustration of orbital structure
of a semibridging carbonyl (CO) ligand.
(Reproduced with permission from reference 124. Copyright 1976 Wiley.)

bonding modes: (1) the mode in which the carbon-bound CO is also π
bonded to a neighboring metal (labeled Π–CO–) center (Structure **3.7d**)
and (2) the more unusual mode in which an additional molecule (or
adduct) is bound to the CO ligand through a σ bond to the oxygen atom
(labeled Σ–CO–) (Structure **3.7g**) (*125–127*). Both bonding modes raise
interesting issues with regard to chemisorption and catalysis.

The CO ligand σ bound through the oxygen atom is known for a variety
of adducts that can accept a lone electron pair, that is, adducts that are good
Lewis acids. The better the Lewis acidity of these adducts, the more likely
they are to form these oxygen-bound CO species. Bound CO ligands of this
nature are known to have adducts containing H^+; alkali atoms such as Li^+,
Na^+, and K^+; adducts containing main-group cations such as Mg^{2+} and Al^{3+};
and adducts containing some early transition metal cations in high oxidation
states such as those of titanium, zirconium, and hafnium (*125*). Alkyl groups
such as a methyl group (–CH_3) are also known to bond to the oxygen atom
of coordinated CO; this bonding can be done with an appropriate alkylating
agent such as CH_3SO_3F.

Because electron-pair acceptors are the species most likely to form the
σ bond to the CO oxygen, it is logical that an enhanced electron-donor
ability of the CO will also enhance the formation of this bonding mode.
This increase in basicity of the oxygen atom occurs naturally as CO is moved
from a terminal site to a twofold bridge site to a threefold site in a complex.
Indeed, this type of adduct formation can lead to the isomerization of a
terminal CO to a bridge CO so that the basicity of the CO oxygen is
adequate for the σ-bonding interaction to occur (*125–127*). The donor
ability of the oxygen atom in CO can also be enhanced in metal complexes
either by placing electron-donor ligands on the transition metal center or
by using a negatively charged carbonyl complex in the preparation of the
species. Because all of these effects are reflected in a shift of the CO
stretching frequency, CO ligands with a stretching frequency greater than
1900 cm^{-1} are unlikely to form these O-bound adducts (*125*).

These O-bonded CO ligands can be quite easily characterized by using vibrational spectroscopy or X-ray crystallography. Formation of these adducts decreases the bound CO stretching frequency by 100–200 cm^{-1} and increases the CO stretching frequencies of the other unbound CO ligands in the complex noticeably. This observation indicates that the adduct bound to the CO oxygen draws a substantial amount of electron density from the metal center and is even known to force other bridging ligands in the complex into a terminal configuration (125–127). Structurally, the O-bound CO bond length is increased by about 0.1 Å, in agreement with the vibrational frequency shift, whereas the M–C bond length is decreased. In addition, the adduct usually forms a C–O–M bond angle that is less than 180°; usually it lies between 162° and 140°; this finding indicates that the hybridization of the oxygen atom of the CO ligand lies between sp^2 and the sp hybridization found in free CO.

The second type of O-bonded CO ligands are those shown in Structure **3.7d**, where the π bonds of the CO molecule donate electron density to an adjacent metal center in addition to the electron donation of the 5σ orbital on the carbon atom (125). This type of bonding can involve participation of one of the π levels, in which case the CO is a four-electron donor, or both of the π levels, in which case the CO is a six-electron donor. Unlike the σ-type O-bonded mode, the π-bonded CO mode can only be characterized crystallographically because its CO stretch frequency lies in the region of conventionally bridged CO ligands.

The π-bonded CO mode generally is found in metal complexes that have large bulky ligands, which form a crowded coordination sphere. This condition leads to a situation in which one CO ligand as a four-electron donor requires less volume than two CO ligands bound as two two-electron donors. The π-bound mode in complexes is usually prepared by allowing a dicarbonyl to stand at room temperature for a few days; this treatment allows one CO to dissociate from the complex so the remaining CO can become a four-electron donor to the metal complex and diminish the crowding in the coordination sphere (125).

These two unusual O-bonded CO coordination modes raise some interesting questions concerning their involvement in catalytic reactions on surfaces. The σ-bonding interaction for CO has already been shown to enhance both the CO insertion reaction into a metal–alkyl bond and CO bond scission in transition metal complexes (125). The geometry of these σ-type oxygen-bound species, however, would seem to be inappropriate for enhancing catalytic activity on flat surfaces, a result that may militate against these O-bonded CO species being operational in coadsorption experiments of CO with alkali (128–130). There may be some reactivity effect noticed on stepped surfaces of aluminum-transition metal alloys, but the question is essentially open. The preparation of the oxygen-bonded CO moiety on surfaces is probably not an impossible task. Trimethylaluminum is a volatile liquid and the most common electron-pair acceptor used for preparing the σ-type O-bonded ligand; it could easily be added to a CO overlayer. Hydrogen atoms prepared from a hot filament may also act as an electron acceptor and lead to the formation of an O-bonded adduct. CF_3 radicals, which can be easily prepared in vacuum (*131*) (*see* page 116), may also form the σ-type O-bonded surface species. The possibility of preparing the π-bonded moiety on surfaces is more difficult to assess because there are no likely approaches that can be considered. It is possible that a π-bonded CO surface mode may form under high pressures of CO, although it would probably disappear under vacuum conditions. The difficulty in characterizing this mode by using vibrational spectroscopy certainly complicates the issue further, although a bonding shift of the 1π level may be detectable with UPS, as was previously reported (*132*).

In addition to these new bonding modes of CO found in metal complexes, a new CO bonding mode was recently reported for a surface system, that is, rhodium dispersed on alumina (*133, 134*). This surface system was the first studied in which more than one CO moiety is found to bond to a single surface metal atom. In the dispersed rhodium system, two (*133*) and even three (*134*) CO molecules have been found, by isotopic labeling studies, to bond to one rhodium center, probably Rh(I). Because this type of bonding is common in organometallic complexes, this result may be viewed as a further bridge between organometallic complexes and surface systems.

3.5.4 Nitric Oxide

Nitric oxide, NO, is probably the most interesting of the diatomics because it not only bonds in the linear terminal (Structure **3.8a**) and bridging (Structure **3.8b**) modes commonly known for CO, but it also bonds in the bent mode (Structure **3.8c**) (*135*). The molecular orbital structure of NO is identical to that for CO except for the presence of a single electron in the π* orbital. Therefore, when NO bonds in a linear fashion, it donates to the metal not only the electron pair in the 5σ level, but also the single π* electron, and formally becomes a three-electron donor. The bent species, on the other hand, is thought to donate only one electron because the octet rule for the nitrogen atom in bent NO can be fulfilled through a two-electron covalent bond with the metal. The most common formalism for describing NO bonding assigns linear bonding to the NO^+ species, which is

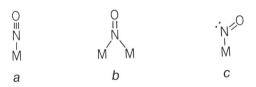

Structures **3.8a–3.8c**. Bonding modes of nitric oxide.

isoelectronic with CO, and assigns bent bonding to the NO^- species. Another more general formalism (*136*) involves keeping track of the total number of electrons in both the metal d orbitals and the π^* orbital of the NO. This record is kept because, after bonding has taken place, both groups of electrons serve a common function in the bonding (discussed later). The general formalism is $\{MNO\}^n$, where M is the metal atom in the complex and n is the total number of electrons in both the d orbitals and the NO π^* level. This formalism can be generalized to any coordinated diatomic ligand.

Substantial interest in NO bonding lies in coming to a better understanding of the nature of bent NO. Several reports (*135-137*) that investigated the bending of NO in detail have been published. From these studies, it is possible to predict which transition metal atom–ligand system will stabilize a bent NO species. The most common class of molecules that can stabilize a bent NO ligand is composed of the square pyramidal complexes of transition metals having eight d electrons (d^8 species). Some of these species are shown in Structures **3.9a** and **3.9b** (*138*). In these structures, the metal atom lies in a plane common to four of the ligands, and the bent NO occupies the apex of the square pyramid of ligands and is bent. If the NO ligand occupies one of the basal sites in the complex, then it is linear (Structure **3.9b**) (*139*). If NO is a ligand in the other five coordinate complexes such as a trigonal bipyramidal complex (Structure **3.9c**) (*140*), then it is usually linear.

Structures **3.9a-3.9c**. Metal complexes illustrating the occurrence (a and b) of the bent and (b and c) linear nitric oxide bonding modes.

Because NO bending is most commonly found in these square pyramidal complexes, the square pyramidal structure is usually the complex studied in theoretical works in order to better understand NO bending (*135-137*). These studies could be significant to surface work because the metal atom environment in a square planar complex resembles the environment of a surface metal atom on a (100) surface on a face-centered cubic metal. Studying the bonding in a square pyramidal complex, therefore, will not only aid in understanding NO bending in general but may also have an extension to surfaces.

A molecular orbital diagram of the interaction of a linear NO ligand with a square planar metal ligand fragment is shown in Figure 3.3 (*137*). The left side shows the relative energies of the d orbitals and how they are split in the crystal field of the four ligands in a square array about the metal atom. The right side of the figure shows relative energies of the n (the 5σ) and π^* orbitals of the NO ligand. The center of the figure shows how these orbitals interact to form bonding orbitals and how their energies change

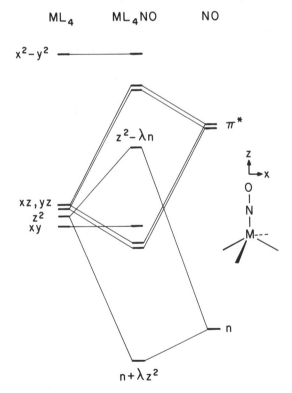

Figure 3.3. Molecular orbital energy diagram for linear nitric oxide interacting with a square planar complex fragment. Occupation of the $Z^2 - \lambda n$ level leads to NO bending.

upon interacting. The electrons in the n orbital of NO fill the lowest bonding orbital, and the remaining d electrons of the metal and π^* electrons of the NO ligand together fill the remaining orbitals; this situation explains the notation $\{MNO\}^n$. This diagram shows that an instability arises when the z^2-λn orbital is filled. It is strongly antibonding, and the molecule will distort to alleviate this instability. The z^2-λn orbital will be filled for $\{MNO\}^8$ species.

When the molecule distorts, it is possible that a reorientation all of its ligands into another five-coordinate structure such as a trigonal bipyramid could occur (*135*). For some cases, however, the instability is alleviated by the bending of the NO ligand. The energy of the z^2-λn level then decreases in energy because the antibonding is relieved. In addition, an interaction between the z^2 level and the π^* level, an interaction that is symmetry forbidden in the linear case, appears (*137*).

In such square planar complexes, the NO ligand is more likely to bend if a judicious choice is made of the other ligands bound to the metal atom. The NO ligand will more likely bend if the z^2 level can be pushed to a higher energy (*137*). This situation will increase the downward slope of the z^2 level in the Walsh diagram of energy versus NO bending angle and thus favor bending (*137*). More importantly, raising the z^2 orbital energy brings it closer in energy to the π^* orbital and so enhances the tendency for the $z^2 + \pi^*$ bonding interaction to occur. The z^2 level can be raised in energy by placing π donor ligands on the metal (*137*). This situation will increase the electron density on the metal and lead to a destabilization of the d orbitals. In bent $\{MNO\}^8$ complexes, the donor ligand most commonly used is Cl^-, shown in Structures **3.9a** and **3.9b**. In the Cl^- ligand, all of the p orbitals are filled and donate electron density to the metal in a π-bonding sense. Although the π donation is not large, it is substantial in comparison with a π acceptor ligand such CO.

When the NO ligand does bend over, it bends in the plane perpendicular to the plane containing the donor ligands. This finding is attributed to the fact that when the NO ligand bends, it must sacrifice one of its d-π^* back bonding interactions while keeping the other. The one that it keeps will logically be the stronger interaction. This interaction is the one in the plane of the donor ligands because the back bonding d orbital in that plane is destabilized relative to the d orbital in the other plane and thus becomes the better π electron donor (*137*).

NO chemisorption has been studied to a fairly large extent and was the subject of a recent review (*141*). The review concluded that on clean transition metal surfaces, nitric oxide generally dissociates at low coverages, whereas at higher coverages, molecular adsorption becomes more favorable resulting in linear, bridged, and possibly dimeric NO species. More recent reports (*142, 143*) have also appeared.

Bent NO is generally characterized conclusively in transition metal complexes by X-ray crystallography (*135*). On surfaces, vibrational frequencies can be used to differentiate between linear and bent NO. For the compound shown in Structure **3.9b**, which has both a linear and a bent NO, vibrational frequencies are detected at 1845 and 1681 cm^{-1}. These frequencies can be attributed to the linear and bent NO stretching frequencies respectively (*139*). In a study of the Cu(110) surface (*144*), peaks at 1564 and 838 cm^{-1} were assigned to bent NO stretching and bending modes, respectively. In addition to the stretching frequency lying near that of bent NO, the presence of the bending mode was considered indicative of diatomic bending because the dipole selection rule would eliminate this mode for a normally bound species.

Studies have also shown that linear and bent NO species may be discriminated on surfaces by the splitting between the O 1*s* and N 1*s* peaks in the ESCA spectrum of NO complexes. In an ESCA study of a variety of NO complexes (*145*), the O 1*s* to N 1*s* splitting was found to be 132 eV ± 1 eV for linear NO structures and 128 eV ± 2 eV for bent NO species. In one report (*10*), this correlation was applied to surface NO overlayers, and the smaller splitting between the N 1*s* and O 1*s* peaks of the overlayers (130 eV for Ni, Pt, and Ir surfaces) was tentatively associated with a bent NO structure.

3.5.5   Dioxygen

Even though it is commonly known that NO can form bent bonded species in transition metal complexes, it is less well known that dioxygen (molecular O_2) can also form bent species such as that shown in Structure **3.10a**. In addition, O_2 not only bonds in a bent fashion, but it also bonds in the side-on mode (Structure **3.10c**) and also in the more common kinked structure (Structure **3.10b**) in which each atom in the diatomic species interacts with the metal atom (*146*) to differing degrees. Dioxygen possesses a molecular orbital structure similar to that of CO, except that dioxygen has unpaired electrons in the π^* orbitals. Dioxygen has little tendency to lose these electrons, but it readily gains either one or two electrons to pair up its two unpaired electrons in the doubly degenerate π^* state. When dioxygen gains one electron, it becomes O_2^- and is called the superoxo species. When it gains two electrons, it is called the peroxo or peroxide species, O_2^{2-} (*146–148*). Generally, the superoxo ligand is bent bonded, whereas the peroxo ligand is side-on bonded (*147, 148*). The two

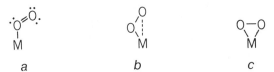

Structures **3.10a-3.10c**. Bonding modes of dioxygen.

species can be easily differentiated by using vibrational spectroscopy because the added electrons in the two species uniquely lower the O–O stretch frequency. For superoxo, the stretching frequency lies between 1075 and 1195 cm^{-1}; for the peroxo species, the O–O stretch frequency lies between 742 and 932 cm^{-1} (*146–148*). X-ray diffraction is also an unambiguous method of determining the dioxygen bonding modes (*148*).

Unlike any of the diatomics discussed earlier, the energy of the π^* levels on dioxygen lies below the energy of the metal d orbitals. For the side-on bound species, therefore, the bond is composed almost completely of d-to-π^* back bonding; very little ligand-to-metal σ donation occurs (*149*).

Currently, there are very few examples of molecular oxygen chemisorbed on single-crystal surfaces (*59*). Apparently, oxygen is so reactive on surfaces that very low temperatures are required to maintain the molecule in the associative state. On Ni (110), O$_2$ dissociates at temperatures as low as 20 K (*59*). Recent studies on Ni(111) have shown, however, that a molecular precursor to chemisorbed atomic oxygen may be stabilized near 5 K (*150*). Associative adsorption of oxygen can be characterized on surfaces by vibrational spectroscopy, the molecule's adsorption behavior, whether or not it desorbs with first-order kinetics and whether or not there is isotopic exchange with desorption (*59*). The most well known case of molecular O$_2$ adsorption is on Pt(111) (*151*); isotope exchange, TDS, and EELS were used to determine the presence of a side-on bonded peroxo species. The presence of an O–O stretch at 870 cm^{-1} was considered indicative of the peroxo species because it falls within the range of frequencies mentioned earlier for peroxo ligands. What was once thought to be adsorption of dioxygen on gold has since been shown to be an interaction of the oxygen molecule with a silicon impurity (*152, 153*).

3.5.6   Acetylene

The final ligand in this series is acetylene (*154*). Although it is not rigorously diatomic, the two triple-bonded carbon atoms have a molecular orbital picture and bonding modes similar to CO and the other diatomics just discussed. The bonding modes include the more commonly known side-on bound species described by the Dewar–Chatt–Duncanson model (Structures **3.11a** and **3.11b**) (*54–56, 155*), and also the lesser known linear alkynyl species (Structure **3.11c**) (*156*), which forms from the acetylide anion ($^-$:C≡C–H). The acetylide species exists in both the terminal and bridging modes. It can also use one or both of its π orbitals as donors and bond in the manner shown in Structure **3.11d**. Acetylene is also known to have bonding modes using several combinations of these σ- and π-bonding interactions (*154, 156*).

The stretching frequency of the C≡C bond in the alkynyl species is heavily dependent on the polarity of the metal–carbon bond. The free anion

Structures 3.11a-3.11d. Bonding modes of acetylene and the alkynyl fragment.

shows the lowest $C\equiv C$ stretch frequency (1992 cm^{-1}), whereas the M–C σ-bonded species shows the highest $C\equiv C$ stretching frequency (2158 cm^{-1}) (156). By analogy with the bonding of the cyanide anion, the vibrational data indicate that the donor orbital on the acetylide anion has a significant C–C antibonding character because the bond order of the $C=C$ bond increases with donation of the lone electron pair. That coordination of the acetylide anion is not known to reduce the $C\equiv C$ stretching frequency indicates there must be essentially no back bonding into the π^* levels of the coordinated acetylide anion (156). Whereas back bonding in coordinated acetylene is common, the negative charge on the acetylide anion appears to push the π^* levels well above the energy level of the d orbitals, an effect that apparently negates a back bonding interaction (104). This situation has been shown to be the case by photoemission studies of metal complexes containing alkynyl ligands (157).

The side-on bound acetylene, which is more common than the alkynyl species, has been studied to a greater extent (155, 158). When bonding to one metal atom, both acetylene π and π^* orbitals have symmetries that match the symmetries of the d orbitals on the metal. This situation leads to the overlaps shown in Figure 3.4. The resulting degree of back bonding is

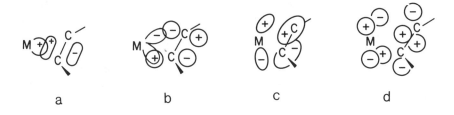

Figure 3.4 Orbital overlaps for acetylene bonding to one metal center. (Reproduced with permission from reference 155. Copyright 1976 Academic Press.)

generally reflected in the C–C bond length (*155*). The interaction of acetylene with two metal atoms, exemplified by the structure of the acetylene hexacarbonyl dicobalt complex (Structure **3.6**), is probably a more realistic structure for comparison with chemisorbed acetylene, however, because it allows increased metal atom coordination about a single chemisorbed molecule. Comparison of the EELS data of acetylene on Ni(111) with corresponding data for the dicobalt complexes (*15*) shows qualitative similarities in the vibrational data; the only differences are attributed to different degrees of hybridization. The hybridization effects were believed to arise because of higher metal coordination on the Ni(111) surface. In a recent valence bond photoemission study of alkynes on Co(0001), bonding similarities were found for substituted acetylenes in alkyne hexacarbonyl dicobalt complexes and on Co(0001) (*11*).

In a theoretical study by Hoffman (*159*), calculations were done to compare the stabilities of the π-bound acetylene bonding mode (Structure **3.11b**) in which the C≡C axis is perpendicular to the M–M axis, with a theoretical di-σ-bound mode, in which the C≡C axis is parallel to the M–M axis (Structure **3.12**). The π-bonded species was more stable by 2.5 eV

Structure **3.12**. Theoretical structure of a di-σ-bound acetylene in acetylene hexacarbonyl dicobalt complex. A metal–metal bond exists between the cobalt atoms.

because all four of its metal–carbon bonding interactions are two-electron stabilizing. Of the four interactions for the di-σ-bonded species, one was repulsive and one was noninteracting. The differences in stability can be visualized by considering, as an analogy, the conversion of tetrahedrane, a structure analogous to the acetylene π-bound mode, to cyclobutadiene, a structure analogous to the di-σ-bound structure (Scheme **3.2**). In this process, the carbon atoms go from unstrained sp^3 hybrids in the tetrahedrane to strained sp^2 hybrids in the cyclobutadiene. A destabilization is thus expected in the process (*159*). In addition, the conversion of the π-bound

Scheme 3.2. Illustration of the theoretical conversion of tetrahedrane to cyclobutadiene.

species to the di-σ-bound species is not directly allowed because the conversion leads to an explicit level crossing of the highest occupied molecular orbital (HOMO) and the lowest unoccupied molecular orbital (LUMO) (*159*). This situation has interesting ramifications for surface-bound acetylene species, because the di-σ-bonding mode has been proposed as a structure for chemisorbed acetylene (*160*).

The side-on bound mode of acetylene has been commonly encountered in surface work and has been the subject of two recent reviews (*11, 53*). The alkynyl species on surfaces, although much less common, has been reported on clean Pd(100) and Pd(111) (*161*) and on oxygen-precovered Ag(110) (*162*) by using EELS. The species is prepared from chemisorbed acetylene, which appears to undergo a single C–H bond scission. On Ag(110), acetylene reacts with the surface oxygen to form OH and the alkynyl species. The alkynyl moiety in both cases is bound terminally through the carbon atom as well as through the π orbitals and results in a species that is almost parallel to the surface.

3.5.7   Generalized Picture of Diatomic Bonding

The diatomic molecules just discussed were grouped together as a class of ligands because they all bond along a reaction coordinate that shows a normal-lying diatomic bending over and dissociating (Structure **3.13a–3.13e**) in a manner that may be similar to a dissociation reaction in a metal complex or on a surface; that is, they bond in either a linear, bent, or side-on fashion. Some of the diatomics can bond in more than one of these bonding modes.

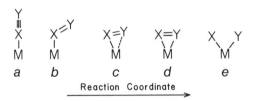

Structures **3.13a–3.13e**. Illustration of the continuum of diatomic bonding modes. Modes include (a) linear, (b) bent, (c) kinked, (c) side-on, and (e) the oxidative addition reaction.

Among the ligands discussed, CN^- bonds linearly, as do N_2 and CO; NO bonds in either the linear or the bent modes; O_2 bonds in the bent or side-on modes; and acetylene bonds in the side-on mode. This situation tends to indicate that there is some key variable within the structure of the diatomics that causes these systematic bonding variations to occur. Hoffman (*104*) has explored the bonding of these diatomic species in general and found that the bonding mode of a given diatomic is determined solely by a series of predictable electronic effects. Through an understanding of these effects, Hoffman was able to propose ways of preparing certain diatomics bound in modes so far unknown for those diatomics. That the bonding of acetylene changes from one end of the scale to the other when it acquires a negative charge in the acetylide anion species can be taken as an indication that the key variable involves the relative energy levels of the ligand and metal orbitals.

In Hoffmann's theoretical treatment (*104*), a variable homonuclear diatomic, $X \equiv Y$, is bound to the apical site of a square pyramidal transition metal–ligand fragment (ML_4), which is identical to the fragment in the study of NO bonding discussed earlier. Calculations were done for the diatomic bound in both linear and side-on modes to correlate the orbital overlaps found at both extremes. A Walsh diagram was subsequently calculated for an M–X–Y bond angle changing from 180° to 90°, the values between the linear and side-on bonding modes. The Walsh diagram is shown in Figure 3.5. From this diagram, it is possible to predict the existence of a given geometry by placing the number of electrons found in the {MXY}" (*see* page

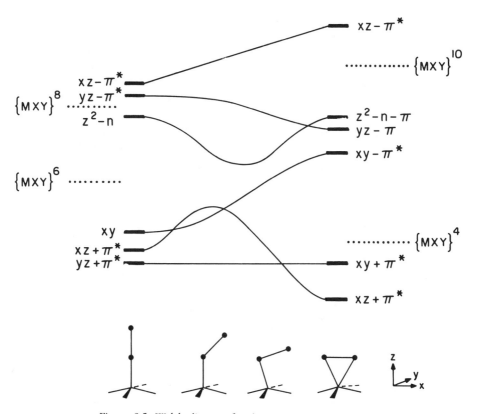

Figure 3.5. Walsh diagram for diatomic bond bending.

38) species into the levels and moving along the reaction coordinate until a minimum in the sum of all the levels is reached. Thus, for $n = 4$ species, a flat-lying species is expected because the lowest two levels are stabilized the most when the diatomic is flat-lying. When $n = 6$, linear bonding is favored because the third lowest level [the xy to $(xy - \pi^*)$] is highly destabilized when the species is flat-lying. The $n = 8$ species is the only one in which a bent or kinked structure is favored because the fourth level [the $(z^2 - n)$ to $(z^2 - n - \pi)$ level], which is the one that is being filled with the seventh and eighth electrons, is stabilized by the bent or kinked structure. For $n = 10$, the flat-lying species is favored, whereas for the $n = 12$ species, all the levels are full and thus the bond order is zero, as is the case for the halogens ($n = 12$).

In general, structural data of actual complexes are in agreement with this treatment (104). When, however, the actual structural data differ from the calculated results, it could be due to two effects. First, the diatomic could be heteronuclear. This situation tends to localize the lower energy orbitals in a bonding–antibonding pair on the less electronegative atom, such as the carbon atom in CO, and disfavor side-on bonding. Second, and most importantly, the electronic structure of the metal center could be modified. This situation could cause the diatomic ligand to be stabilized in a different bonding mode. Because in transition metal complexes it is possible to modify the electronic structure of the metal center by carefully selecting ligands with specific electron-donor or electron-acceptor properties, Hoffman has qualitatively described the nature of these effects (104). These effects can also be qualitatively applied to the surface bonding of diatomics and thus may lead to new diatomic bonding modes on surfaces.

The interesting electronic effects center predominantly around the nature of the bent diatomic species because it would be most interesting to make a linear species such as CO bend over or to make a bent species such as O_2 linear. Qualitatively, the bending process occurs because, upon bending, as discussed for NO, the unstable $(z^2 - n)$ level can be stabilized by interaction with the $xz + \pi^*$ level. This situation leads to an interaction of the z^2 level on the metal with the π^* level on the diatomic, as shown in Figure 3.6, and a bonding geometry that is more stable than that for the corresponding linear mode. By bringing the energy of the d_z^2 level on the metal closer to the energy of the π^* level in the diatomic, the propensity for this bending should be enhanced (104).

From this description, it would seem that a linear diatomic that has no known bent modes, such as CO, N_2, or CN^-, could be made to bend over by bringing the d_z^2 level of the complex to an energy that is close to the respective energies of the π^* levels of the diatomic species. Because these π^* levels all lie to higher energy relative to the metal d_z^2 level, the d_z^2 level would have to be destabilized for the bending process to become possible. This destabilization could be done, theoretically, by placing on the metal

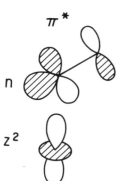

Figure 3.6. Orbital overlap for the bent bonding interaction. Bonding results from interaction of metal d_{z^2} and one diatomic π^* orbital.

center donor ligands that increase the electron density on the metal and thus destabilize all the metal orbitals, the d_{z^2} included. This effect has already been discussed for the mechanism of NO bending (*137*). Of the three possible diatomics, the CO ligand has the lowest energy π^* level, and thus it is the most likely linear-bound diatomic to be prepared in the bent mode.

Arguments for the preparation of a linear-bonded dioxygen ligand follow a pattern similar to those presented for diatomic bending (*104*). For dioxygen, the π^* levels are so low in energy relative to those of CO that they lie very close in energy and, in fact, slightly below the energy of the d_{z^2} level in transition metal complexes. This condition gives the dioxygen ligand a propensity to bend that is so strong that no linear dioxygen ligands have as yet been characterized. The best way to prepare a linear O_2 ligand, therefore, would be to move the energy level of the d_{z^2} level far away from the energy of the π^* level of dioxygen in order to break the favored $d_{z^2} - \pi^*$ interaction. Because the π^* level lies below the d_{z^2} level, destabilizing the d_{z^2} level could break the $d_{z^2} - \pi^*$ interaction and form a linear dioxygen ligand (*104*).

Although these arguments are generally straightforward theoretically, it is very difficult to implement these ideas experimentally in organometallic complexes. It is not usually possible to prepare complexes with any particular combination of ligands desired. However, on surfaces, preparing a system of this nature is quite easy by simply coadsorbing with chemisorbed CO or dioxygen, donor species that may serve to destabilize the d_{z^2} levels in the surface atoms. A logical example for this process is the much studied coadsorption of alkali atoms (lithium, sodium, and potassium) with CO or NO on transition metal surfaces (*128*). Alkali atoms are excellent electron donors and may serve a function on diatomic-covered surfaces that is similar to the function of donor ligands such as Cl⁻ in complexes containing bent NO ligands. It is, therefore, possible that coadsorbed alkali atoms cause chemisorbed diatomic species on surfaces to bend over.

The data so far collected on these alkali–diatomic systems are completely analogous to the arguments presented for diatomic bending in organometallic complexes. CO coadsorption with potassium atoms (*129*) leads to substantial shifts from terminal to bridge-bonded CO and also to a continual decrease in the CO stretching frequencies with potassium coverage. These effects are attributed to a substantial charge donation from the potassium through the metal substrate into the π^* level of the CO, an argument completely analogous to the effect seen in organometallic complexes. A recent calculation (*129*) indicated that the coadsorbed potassium can decrease the energy splitting between the surface bands and the $2\pi^*$ level of the CO. Again, this is precisely that proposed by Hoffmann to lead to CO bond bending in complexes. Indeed, the coadsorption of potassium with chemisorbed NO, a diatomic species having a great propensity for bending, leads to enhanced NO dissociation on Pt(111) (*163-165*). This result could indicate an alkali-enhanced NO bending, as proposed for metal complexes, so that the oxygen end of the molecule can interact with the surface. Most importantly, results have now been reported that show that CO coadsorbed with alkali species does exhibit characteristics consistent with the CO moiety being tilted on the surface (*132, 166*). These results include isotopic scrambling of C and O atoms in labeled CO coadsorbed with potassium (*167a, 167b*).

3.5.8   Sulfur Analogues of CO and NO

In addition to the diatomics discussed earlier, there are two diatomics that have not been studied on surfaces because they are very difficult to prepare. They are the thio (sulfur) analogues of CO and NO: thiocarbonyl (CS), and thionitrosyl (NS).

Thiocarbonyl bonds to transition metal centers in the same manner as CO, that is, through the terminal (Structure **3.14a**) and bridge-bonded (Structure **3.14b**) species. In addition, the end-to-end bridge species in Structure **3.14c** is also known. The terminal structure is, however, the most common (*168a, 168b*). Thiocarbonyl is a very interesting ligand because it is both a better σ donor and a better π acceptor than CO. The 7σ lone-pair donor orbital on CS has a higher energy than the 5σ level on CO, and the 3π acceptor orbital on CS is lower in energy than the 2π level on CO (*168a, 168b*). The lower 3π level also indicates that CS probably has a propensity

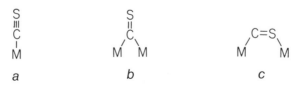

Structures **3.14a-3.14c**. Bonding modes of thiocarbonyl (CS).

to bend that exceeds CO and may yield bent structures more easily than the CO ligand, according to the earlier discussion of atomic bending.

Thiocarbonyl can be characterized much like CO; it has terminal and bridging CS stretching frequencies in the ranges of 1409 to 1161 cm^{-1} and 1160 to 1106 cm^{-1}, respectively (*168b*). In Raman spectroscopy, a high polarizability is expected but not seen, probably because its change in polarizability is minimized by coupling of the π^* and π levels through the metal atom (*168b*).

The problem in working with thiocarbonyl is that it is difficult to prepare and impossible to store in pure form (*169*). Most of the complexes of thiocarbonyl are prepared by using more stable precursors such as carbon disulfide (CS_2) and thiophosgene Cl_2CS which undergo bond scission and release the CS; these processes leave the excess chlorine or sulfur atoms behind to be removed by purification (*168a*). Most purification procedures, however, cannot be carried out on surfaces, and a method must be found that uses CS gas directly. CS was originally prepared by passing CS_2 through a high voltage ac discharge (*170*). The CS was reported to be stable for short periods in the gas phase but polymerized readily as a liquid and on the surfaces of its container. If the CS could be separated from the CS_2 by some type of fractional distillation process (*169*) in which the CS does not condense, then it is possible that the small quantities required for chemisorption work could be prepared and used immediately. The ac discharge method was originally used as a preparative procedure for CS chemistry (*171*); no attempt to separate out the CS_2 was made. Recently, partial CS purification has been achieved by passing the CS–CS_2 mixture through a trap cooled to –112° C (*169*).

In a recent study of Pt(111) (*100b*), methanethiol (H_3CSH) was adsorbed on the surface as a means of preparing C≡S by a hydrogen-stripping reaction. Although some CS was detected by HREELS, it could not be prepared in pure form.

The final diatomic species to be discussed is thionitrosyl (NS), the sulfur analogue of NO. Complexes of NS have been known only since 1974 (*172*). It too has bonding modes analogous to its oxide analogue; terminal (Structure **3.15a**), bridging (Structure **3.15b**), and bent (Structure **3.15c**) modes are known. Like CS, it is a better σ donor and a better π acceptor than its oxide analogue, and it can be characterized by using vibrational spectroscopy because terminal NS has its stretching frequency between 1150

Structures **3.15a-3.15c**. Bonding modes of thionitrosyl (NS).

and 1370 cm^{-1}. Bent NS has its stretching frequency between 1110 and 1130 cm^{-1} (*172*). NS has also been detected in the ac discharge when the reacting gases were N_2 and sulfur vapor (*173*). The lifetime of NS is so short, however, that it probably could not be trapped out on a surface.

3.6   Substituted Diatomics

The third class of ligands to be described are the substituted diatomics, which are species that have a part resembling a diatomic ligand and an additional substituent attached to either one or both atoms of the diatomic part. The bonding modes of these ligands in metal complexes are very much like the bonding of diatomic species. The potential advantage of studying this type of ligand is that the substituent can provide additional understanding of the coordination bond because the substituent's geometry in the bound ligand leads to conclusions concerning both rehybridization and back bonding in the diatomic part of the molecule. In addition, the substituent moiety can be selected to be either an electron donor or an electron acceptor, a choice that can either raise or lower the energy levels of the diatomic moiety and result in a change in the metal–diatomic bonding interaction (*174*).

The advantage of additional understanding of the coordination bond can be visualized by considering how the respective hybridizations of each atom in the diatomic change with changing back bonding. Structure **3.16a** shows a valence bond picture of terminally coordinated CO with no back bonding. This same picture applies to a CO ligand that has a back bonding interaction that is equally shared by both of the π^* levels on CO. In this case, both the carbon and the oxygen atom are *sp* hybridized. Structure **3.16b** shows a valence bond picture of CO undergoing back bonding through only one of its π^* levels so that a metal–carbon double bond results. Although in this case, the carbon atom maintains an *sp* hybridization, the oxygen atom becomes *sp*^{2} hybridized. It would be very interesting to determine if chemisorbed CO ever bonds in the mode shown in Structure **3.15b**. Unfortunately, however, no concomitant structural change occurs when the oxygen atom is rehybridized. Thus, there is no way of determining if this rehybridization actually occurs, or if the back bonding occurs through the equal use of both π^* levels. Since substituted diatomics have something attached to that outer atom (although not for the CO species itself), the

Structures **3.16a–3.16c**. Illustration of the effect due to anisotropic back bonding in a diatomic ligand and a substituted diatomic ligand. For both cases, the outer atom of the diatomic becomes *sp*^{2} hybridized; structural evidence for this rehybridization is evident only for the substituted diatomic.

hybridization of the second atom can be specified by determining the angle θ the substituent makes with the diatomic part of the ligand (Structure **3.16c**). An angle of 180° indicates that no rehybridization of the Y atom occurs, and so back bonding occurs equally through both π^* levels. An angle that is close to 120° indicates significant rehybridization due to the unequal or anisotropic back bonding through only one of the π^* levels. From the discussion of substituted diatomics to follow it seems that both bonding modes are possible depending on the system being studied (*51, 174*).

The second advantage of substituted diatomics is that changes in the energies of the bonding orbitals on the diatomic can be made by an appropriate choice of substituent. Electronegative species can, through an inductive effect, lower the energy levels of the orbitals on the ligand, whereas electropositive species can raise the energy levels. Although the d-to-π^* back bonding in diatomic coordination could only be controlled by raising or lowering the d-orbital energies on the metal center through ligand donors or ligand acceptors, with substituted diatomics, electronic energy variations on the ligand as well as on the metal can be varied (*174*).

3.6.1 Substituted Cyanides

One of the most commonly available substituted diatomics is acetonitrile or methyl cyanide ($H_3C-C\equiv N$), which is the simplest of the class of molecules called the nitriles (R–CN), for which the variable substituent R can be almost any kind of organic substituent. Acetonitrile is a common solvent and it and most any other nitrile species are satisfactorily volatile for chemisorption work (*175*). Acetonitrile bonds to metal centers in two possible modes: (1) the more common linear mode (Structure **3.17a**) and (2) the very rare side-on mode (Structure **3.17b**) (*174*), which has only been observed for one platinum complex (*176*). The $C\equiv N$ stretching frequency of acetonitrile is near 2200 cm^{-1} and increases with linear coordination by about 30 cm^{-1}. This increase indicates that the donor electron pair is $C\equiv N$ antibonding and that little back donation is occurring into the π^* levels of the CN moiety (*174*). For some low-valent complexes of acetonitrile, the frequency shifts to lower energies. This shift occurs because the lower oxidation state of the metal sufficiently destabilizes the d orbitals and allows some d-to-π^* bonding to take place. For the one documented case of side-on coordinated

Structures **3.17a-3.17d**. (a–c) Bonding modes for substituted cyanides. (d) Surface bonding mode of acetonitrile on Pt(111).

acetonitrile, the C≡N stretch falls at 1734 cm^{-1}. This finding indicates a substantial back bonding interaction (*174*).

A special case of substituted cyanide is the cyanide dimer or cyanogen. When it does not undergo carbon–carbon bond cleavage, it bonds as a typical nitrile, although both ends can coordinate independently to separate metal atoms. The result is a linear species that has no obvious relevance to surface chemisorbed structures (Structure **3.17c**).

Acetonitrile has been studied on several surfaces, and the limited data tend to indicate bonding modes similar to those found in metal complexes. On Ni(111) (*177a*), the molecule was observed by TDS to be weakly and reversibly bound, and normal coordination has been proposed for the bonding mode. HREELS data, however, gave a C≡N stretching frequency that fell in the region characteristic of bridging acetonitrile (*177b*). On Ni(100) (*177a*), acetonitrile was also weakly bound, except that the desorption maximum occurred 20° higher in temperature than it did on Ni(111). Acetonitrile was also found to be weakly bound to dispersed rhodium (*178*). On Ni(110) (*177a*), decomposition occurred above 400 K. This result indicates a stronger interaction of acetonitrile on the (110) surface. On the stepped Ni[9(111) × (111)] surface (*65*), acetonitrile underwent 10% decomposition through C–H scission. On Pt(111) (*17*), the C≡N stretching frequency was found at 1615 cm^{-1}. The substantial reduction in the stretching frequency, in comparison with both linear-bonded and physisorbed acetonitrile, clearly indicated a side-bonded species. But the stretch frequency was much lower than that of the side-on π-bonded species in the metal complex, and because the frequency exactly matched that of the C=N double bond, a di-σ-bound species (Structure **3.17d**) was proposed for the acetonitrile structure on Pt(111) (*17*).

3.6.2   *Substituted Isocyanides*

If the alkyl group is attached to the nitrogen atom instead of the carbon atom of the cyanide diatomic, then the isocyanide (or isonitrile) group of ligands is obtained. These species bond in the linear mode (Structure **3.18a**) (*174*), the C–N bridge mode (Structure **3.18b**), and the combined σ–π bridging mode (Structure **3.18c**) (*179*). Isocyanides are also known to coordinate in a bridging mode in which the methyl isocyanide ligand is bent (Structure **3.18d**) (*178, 180*). Terminally bound isocyanides can also bond in the bent configuration (Structure **3.18e**) ($\theta = 135°$) (*181*) although such bonding is rare. This situation indicates that for the isocyanides, the back bonding can be isotropic or anistropic, depending on the electronic structure of the isocyanide complex. A concise understanding of these back-bonding effects is, however, not yet available (*179*).

Similar to the stretching frequencies of the cyanide anion and acetonitrile, the CN stretching frequency may increase when σ bonding to the metal

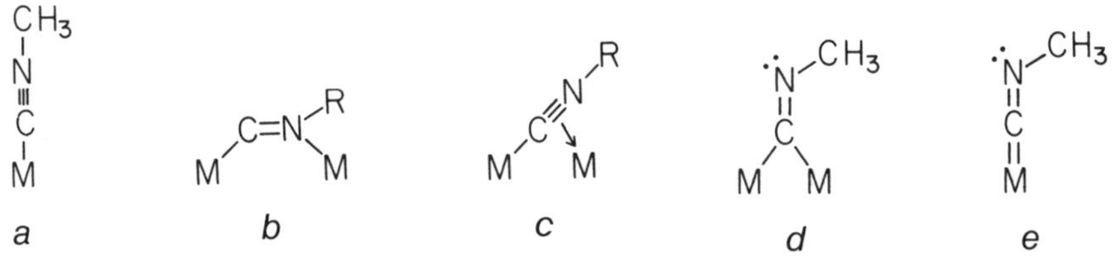

Structures **3.18a-3.18e**. Bonding modes of alkyl isocyanides.

atom predominates and may decrease when π back bonding is significant. Relative to CO, methyl isocyanide is a better σ donor and a poorer π acceptor because the lone-pair donor orbital (the 5σ) and the π acceptor orbital (the 2π) are both higher in energy than the corresponding CO levels (*174*). The reduced π-acceptor ability can explain the unusual shifting in the CN stretch energy because a small amount of *d*-orbital stabilization can significantly widen the energy gap between the *d* and π^* levels and eliminate the back bonding interaction.

On various low-index planes of nickel, methyl isocyanide was found to be strongly bound (*177a, 177b*) and in TDS experiments, decomposition prevailed over desorption. On carbon-precovered nickel surfaces, the isonitrile isomerized to acetonitrile at 90°C (*177a*). On Ni(111), the C≡N stretching frequency of methyl isocyanide fell in the region characteristic of bridging isocyanide (*177b*). On dispersed rhodium (*178*), the bonding of methyl isocyanide was found to be analogous to that of CO; no dissociation or isomerization was detected. On Rh(111) (*182a*) nondissociative adsorption occurred at bridging sites for low coverages at 120 K, whereas terminal sites became favored at higher coverages. Above 350 K, the methyl isocyanide decomposed to H_2, HCN, and N_2. On Ag(311) (*182b*), HREELS and ESDIAD were used to show that methyl isocyanide can bond to the surface in modes that are either perpendicular or tilted relative to the macroscopic crystal surface. Dynamical LEED studies of bridging sites of adsorbed methyl isocyanide would be a good method of determining whether the surface adsorbate assumes the linear or bent structures known for metal complexes of isocyanides. Because the CO molecule bonds in a similar fashion, much insight could be gained into the nature of CO back bonding by determining the anisotropy of methyl isocyanide back bonding.

3.6.3 Substituted Dinitrogen

The cyanide anion is not the only diatomic species that can have attached substituents; dinitrogen can also bond to substituents. When a substituent is located at both ends of the dinitrogen ligand, diazene species result (Structure **3.19a**), which bond in the side-on mode of coordination, much like that of ethylene (discussed later). The R substituent is usually a phenyl group. When atomic radii are taken into account, the M–N bonds are shorter and the N–N bonds are longer than the corresponding substituted ethylene species. This finding indicates that diazenes are better back bonders than ethylene ligands (*54*).

$$
\begin{array}{c}
R \diagdown N = N \diagup R \\
\diagdown \diagup \\
M
\end{array}
\qquad
\begin{array}{c}
R \diagdown C = C \diagup R \\
\diagdown \diagup \\
M
\end{array}
$$

a   b

Structures **3.19a** and **3.19b.**   Side-on bonding mode for disubstituted dinitrogen and disubstituted acetylene.

When only one nitrogen atom of dinitrogen has a substituent, the diazo ligand group results. Diazo species such as diazomethane are most commonly used to prepare carbenes (CR_2 ligands), because the N_2 moiety can so easily dissociate from the CR_2 part of the molecule (*92*). If, however, mild conditions are used in preparing the complex, it is possible to prepare stable species containing diazo ligands in any one of a variety of bonding modes (Structures **3.20a-3.20f**) (*51, 92, 183*). The CR_2 group can also be replaced by a phenyl substituent (*50*). Diazomethane is a marginally stable liquid and has been investigated in two surface chemisorption studies (*72, 184*).

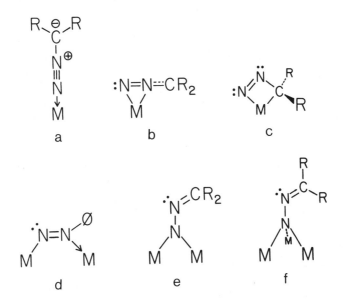

Structures **3.20a-3.20f.** Bonding modes for monosubstituted dinitrogen species (diazo species).

What is most interesting about the diazo group is that for the phenyl-substituted diazo ligand (phenyl =⬡, designated ϕ), the N–N–C bond angle θ shown in Structure 3.21a is a function of the magnitude of the anisotropy of the d to π^* back bonding (*51*). This finding indicates that for the phenyl-substituted diazo ligand (called the aryldiazenato species), the valence bond picture in which the back bonding is anisotropic is the most accurate because the phenyl diazo ligand is always bent (*51*). This anisotropy is probably due to a conjugation interaction between the benzene ring and the doubly bonded diazo part of the ligand. This kind of stabilization is quite common in organic chemistry (*185*) and cannot be achieved unless the central nitrogen atom has a π-type overlap between a p orbital on the nitrogen atom and a p orbital on the benzene ring. This condition is most readily accomplished if the nitrogen atom adjacent to the ring is sp^2 hybridized. Although this structure offers little insight into the nature of metal–diatomic interactions, it shows that anisotropic back bonding is known when other ligand stabilization pathways are available. The phenyl-substituted diazo species can also bond to a metal center in the manner shown in Structure 3.21b, where both nitrogen atoms are sp^2 hybridized and result in a ligand that is doubly bent.

a $\qquad\qquad$ b

Structures 3.21a and 3.21b. Bent bonding modes of the phenyl diazo ligand.

Characterizing phenyl–diazo ligands is relatively easy because the N–N stretching frequency falls as a monotonic function of bending angle between 2100 and 1600 cm^{-1} (*51*). Preparation of surface-bound diazo species is more complicated. For diazomethane chemisorption, the volatile liquid can be used, and if it is placed on a low-reactivity surface, scission may not occur. The more interesting phenyl–diazo species, by contrast, is not volatile, so alternate routes are necessary for its preparation on surfaces. In metal complex synthesis, some success has been found by reacting a coordinated NO ligand with aniline (H$_2$N–ϕ) (Scheme 3.3a) to generate the phenyl–diazo ligand and H$_2$O (*186*). In addition, the reaction of phenylhydrazine (H$_2$N–NHϕ) with metal complexes (Scheme 3.3b) has also yielded the aryl–diazenato ligand (*187*). In both cases, all of the reactants are volatile and thus could be brought to a surface. Experimental studies are necessary to determine whether or not the interesting aryldiazenato ligand can be prepared on a surface so that its bending phenomena can be studied on a surface.

Schemes 3.3a and 3.3b. Two known organometallic synthetic pathways to the phenyl diazo ligand in metal complexes. Both are carried out with volatile species and thus may yield similar results on surfaces.

3.6.4   Substituted Acetylene

Acetylene, although it is not rigorously diatomic, can also undergo substitution of its hydrogens with almost any other organic species (Structure **3.19b**). However, regardless of the substituent placed on the ligand, the degree of rehybridization will be changed according to the degree of electronegativity of the substituent group. With increasing electronegativity, the π^* levels are stabilized and allow enhanced back bonding and thus increased rehybridization. Therefore, if the R group is strongly electron withdrawing (for instance, $R = CF_3$), the metal–ligand bond will be stronger than if the R-group is electron releasing ($R = CH_3$). Excellent candidates for chemisorption are the haloacetylenes (*188*), that is, Cl-C≡C-Cl. They should bond very strongly to the surface and may undergo some very interesting surface chemistry.

3.6.5   The Formyl Ligand

A substituted diatomic with potential importance in methanation chemistry is the formyl ligand (Structure **3.22a**). The formyl ligand is simply a CO ligand with a hydrogen atom bound to the carbon atom. It is usually characterized by its CO stretching frequency, which is of a medium intensity

a   b

Structures **3.22a** and **3.22b**. Illustration of the (a) formyl and (b) formimidine ligands. Both can be viewed as CO or methyl isocyanide ligands with a hydrogen atom bound to the coordinated carbon atom.

relative to the CO stretch of carbon monoxide and which falls between 1530 and 1630 cm^{-1} The C–H stretch between 2830 and 2695 cm^{-1} is, however, not usually intense enough to be detected (*189*). It is generally stabilized by the presence of good donor ligands in the complex (*189*), those same ligands that lead to diatomic bending and enhance the possibility that CO bending is involved in the methanation process.

Preparing this ligand on a surface is a difficult challenge. In one attempt on dispersed rhodium (*190*), formaldehyde, glyoxal, coadsorbed H$_2$ and CO, and atomic hydrogen bombardment of chemisorbed CO were employed in the hope that they would produce the formyl species on a surface. These methods involved hydrogen capping, use of dimers, and surface reactions—three of the four surface preparative techniques discussed earlier in this chapter. None were successful. It is possible, however, that the coadsorption of a donor species such as potassium may facilitate the formation of a formyl species, as is apparently the case in metal complexes. Potassium, as discussed in the section on the generalized picture of diatomic bonding, may enhance the tendency for CO to bend over and increase its reactivity to surface hydrogen. The thioformyl ligand (HCS) is probably more likely to form on a surface than HCO because hydride migration to CS occurs readily in complexes and, therefore, may also occur readily on a surface (*189*). Hydride migration to the coordinated carbon atom in isocyanide ligands is also facile (rigorously known as isocyanide insertion into a metal hydride bond) in metal complexes and results in the formation of the formimidine ligand (Structure **3.22b**) (*179, 191*). This species is also likely to form on surfaces and has already been proposed as a model intermediate in the methanation reaction (*191*).

3.7 Ethylene and the Cumulenes

One class of ligand that has seen little emphasis in chemisorption is composed of the cumulenes, which are linear triatomic species in which the central atom is bound to both outer atoms through double bonds (Figure 3.7). The bonding mode is accomplished by *sp* hybridization of the central atom, which is then able to form two π-type interactions. The outer atoms are *sp^2* hybridized so that each has a single *p* orbital available for π bonding. Because the bonding interaction of the cumulenes to metal centers is so similar to that of ethylene, ethylene is also discussed in this section. The cumulenes include species such as allene (H$_2$C=C=CH$_2$),

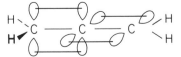

Figure 3.7. π-bonding interactions in the allene molecule; these interactions are common for cumulene species.

carbon dioxide (O=C=O), carbon disulfide (S=C=S), carbonyl sulfide (O=C=S), ketene ($H_2C=C=O$), and sulfur dioxide (SO_2).

3.7.1 Ethylene

The coordination bonding of the cumulenes is very much like that of ethylene (*54*) and is described by the well-known Dewar–Chatt–Duncanson model of ethylene coordination (*55, 56*). This model involves the donation of the electron density in the π bond on the ethylene molecule to the metal and a back donation of *d*-orbital electron density to the π^* level of the ethylene moiety (Figure 3.8). The resulting bonding mode is shown in Structure **3.23a**, in which ethylene coordinates to only one metal center (*54*). Similar to the diatomics, the relative energies of the metal *d* orbitals and the π and π^* levels on the ethylene molecule can lead to changes in the degree of the donation and back donation between the ethylene molecule and the metal atom in the complex. For ethylene, these changes in the bonding are manifested by changes in the carbon–carbon and metal–carbon bond lengths, the degree of C–H bond out-of-plane bending, and various asymmetries in structure. These structural asymmetries include sliding, for which the centroid of the ethylene ligand is displaced from the mirror symmetry plane of the complex; torsional twisting, for which the

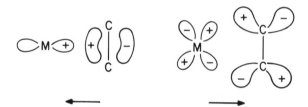

Figure 3.8. Orbital overlaps illustrating the Dewar-Chatt-Duncanson mode or π-bonding mode of ethylene coordination. (Reproduced with permission from reference 54. Copyright 1976 Academic Press.)

a	b	c

Structures **3.23a-3.23c**. (a) Ethylene bonding mode, (b) allene side-on bonding mode, and (c) structure of methylene cyclopropane whose IR spectrum is similar to that of coordinated allene.

out-of-plane bending of different substituents on the ethylene ligand is of differing magnitudes; and rotations of the olefin ligand out of the plane of the metal complex (*54*).

The first ionization potentials (IP) of various metal atoms that form complexes with ethylene are proportional to the carbon–carbon bond length of the bound ethylene molecule, that is, the lower the ionization potential, the longer the bond. Because the first IP measures the energy required to remove one electron in the HOMO to infinity, structures with low IP have destabilized *d* orbitals (*54*). This situation enhances the back bonding interaction. The added electron density in the π^* level further weakens the carbon–carbon bond and increases the carbon–carbon bond length.

Similar to the case for substituted diatomics, substituents on the ethylene ligand can affect ethylene coordination to a metal center through changes in the ligand's electronic structure. The metal–carbon bond length decreases and the of out-of-plane bending increases as the hydrogen atoms on the ethylene molecule are substituted for other organic substituents, such as the $-C{\equiv}N$ moiety or the halogens. In general, the increased electron-withdrawing effect of the substituent inductively stabilizes the π^* level of ethylene and enhances the back bonding. These occurrences result in the geometric changes just described (*54*). When an ethylene molecule has different types of substituents on its carbon atoms, structural asymmetries result. For a species such as $Cl_2C{=}C(CN)_2$ (1,1-dichloro-2,2-dicyanoethylene), the two metal–carbon bond lengths differ by 0.1 Å (the CCl_2 side has the shorter M–C bond length). This finding reflects the higher electron acceptor ability of the chlorine substituents. The same species also undergoes a sliding effect to bring the $C{\equiv}N$ substituents closer to the metal atom. This effect, however, is not well-understood (*54*).

Ethylene has been studied on many surfaces, and its chemisorption behavior has been the subject of two recent reviews (*11, 53*). Both photoemission (*192*) and HREELS (*193*) studies have shown that chemisorbed ethylene undergoes both out-of-plane bending and carbon–carbon bond lengthening, analogous to ethylene bound in transition metal complexes. These and other results are thought to validate the Dewar–Chatt–Duncanson model (*55, 56*) for ethylene bonding on surfaces. LEED work indicates, however, that unlike transition metal complexes, the ethylene molecule sits in a threefold hollow site (*194*) in an attempt to maximize its metal atom coordination. Surface studies have not, however, been extended to substituted ethylenes such as the tetrachloro-, tetrafluoro-, or tetracyanoethylenes or to any number of species that have carbon atoms of greatly differing back bonding ability. Surface studies of these substituted ethylene species may lead to a new chemistry and new species on surfaces.

Even though ethylidyne is usually closely associated with ethylene chemisorption, in this work ethylidyne is discussed under the class of fragment ligands of hydrocarbons and thus will not be discussed here.

3.7.2   Allene

Although the cumulenes have been studied much less extensively in coordination complexes than ethylene, they all bond to the metal atom through one of their double bonds much like ethylene. This situation is shown for allene in Structure **3.23b**. By analogy to the rehybridization in ethylene, the bending angle θ in the coordinated cumulene species is proportional to the length of the coordinated carbon–carbon bond, which in turn is dependent on the degree of back bonding (*155*). The carbon–carbon bond length of the coordinated π bond is longer than that of the uncoordinated π bond. The metal–carbon bond to the central carbon atom is shorter than the metal–carbon bond to the outer carbon atom (*155*). IR studies have shown that both carbon–carbon stretch frequencies lie between 1650 and 1910 cm^{-1}. For $X_2Pt(C_3H_4)$ (X = halogen), because the asymmetric stretch frequency for the coordinated allene ligand lies at 1680 cm^{-1}, a greatly rehybridized allene ligand is indicated. This conclusion is made because the vibrational frequency lies close to that of methylene cyclopropane (Structure **3.23c**).

Few chemisorption studies of allene gas are known (*195*). It may, however, be an interesting candidate for dynamical LEED work because the bending angle, a measure of rehybridization, can be determined. Because LEED cannot conveniently detect hydrogen in chemisorbed ethylene, the position of the uncoordinated carbon atom can provide one of the few detectable measures of olefinic rehybridization on a surface.

3.7.3   Carbon Disulfide

A second cumulene is carbon disulfide (S=C=S). Carbon disulfide can bond like allene (Structure **3.24a**) (*196*) or, more rarely, it can bond through a sulfur lone pair maintaining its linearity (Structure **3.24b**). It can also bridge in a number of ways as exemplified by Structure **3.24c** (*168*), but because the ligand lies between the metal centers in these modes, the modes have no obvious analogy to surface bonding.

The structure of the bound cumulene is very similar to that of allene. The coordinated C–S bond is longer than the noncoordinated bond, and

Structures **3.24a-3.24c**. Bonding modes of carbon disulfide.

the metal–carbon bond is noticeably shorter than the metal sulfur bond (*168b*). Both C–S bonds are longer than those in the uncoordinated molecule. This situation indicates that antibonding orbitals must be involved in the bonding (*197*). The vibrational spectra show two carbon–sulfur stretch peaks: (1) the coordinated C–S bond stretch (in-ring) which lies between 623 and 653 cm^{-1} and (2) the other C–S bond stretch (non-ring), which lies between 955 and 1235 cm^{-1}. The asymmetric stretch for the linear bound CS_2 (Structure **3.24b**) lies at 1500 cm^{-1}, 20 cm^{-1} below that of gas-phase CS_2 (*168b*). Currently, no reports have been made of the chemisorption of CS_2, a volatile liquid. CS_2 may be interesting to study as an adsorbed species for a reason similar to that for allene; the linear species would also be interesting to study by using angle-resolved UPS to determine if it lies normal to the surface (*198*).

3.7.4   Heterocumulenes

A third type of cumulene species is carbonyl sulfide (O=C=S), which is the most common ligand in a class of heterocumulenes that includes S=C=NR, O=C=NR, and S=C=Se (*197*). These species presumably coordinate to the metal center in a complex in a manner similar to the side-on bonding mode of allene and carbon disulfide. The interaction, however, is apparently so strong that scission occurs and results in a bound diatomic species. For the OCS ligand, coordination yields bound CO; the sulfur atom is removed in the purification procedure (*196*). For carbonyl sulfide, the scission reaction occurs exclusively through the C=S bond. For S=C=NR, a similar result—coordinated alkyl isocyanide—is obtained. When Se=C=S is the reacting ligand, the thiocarbonyl ligand results, and the selenium atom is removed in purification. This reaction is unusual because it does not occur for species in which the outer atoms in the molecule are alike. Little is known about the mechanism of this reaction, but it probably involves the outer uncoordinated atom, which affects the bond strength of the in-ring bond (*196*).

This unusual scission reaction does have an analogous reaction on surfaces (*199*). On both nickel and tungsten films, thiocarbonyl adsorbs associatively at 195 K but dissociates at higher temperatures into CO, which desorbs from the surface, and sulfur, which remains behind. This reaction of carbonyl sulfide could be very advantageous in surface studies because the use of carbonyl sulfide could provide a method for coadsorbing a sulfur atom near a CO molecule in a reproducible manner. If the scission reaction could be demonstrated to occur generally on a surface for COS gas, it may provide a method for measuring the CO-sulfur interaction under stoichiometric conditions because the CO-to-S surface composition ratio is necessarily 1:1.

Ketenes, $R_2C=C=O$, also undergo double-bond scission during reaction with metal complexes, but they can be cleaved at either double bond. When the C=O bond is cleaved, a bridging carbene results (Structure **3.25a**) (*200*), and the oxygen atom combines with a CO ligand to form CO_2, which subsequently departs from the molecule. If the C=C bond is cleaved, a carbene and a CO ligand result (Structure **3.25b**) (*183*), both bound to the same metal atom. However, thioketene, $R_2C=C=S$, bonds like CS_2 and undergoes no scission processes (*197*).

Structures **3.25a** and **3.25b**. Ligand species resulting from bond scission in ketene occurring between (a) the carbon and oxygen atoms and (b) the two carbon atoms.

3.7.5   Carbon Dioxide

A cumulene species that has been studied on surfaces is carbon dioxide (O=C=O). In transition metal complexes, CO_2 bonds much more weakly than carbon disulfide. Thus, there are fewer known CO_2 complexes than CS_2 complexes. CO_2 bonds primarily by the side-bound (Structure **3.26a**) and end bound (Structure **3.26c**) structures common to the cumulene ligand class, but an additional mode is known for CO_2 where bonding occurs through the carbon atom (Structure **3.26b**) (*201*). Athough carbon disulfide bonds to almost every transition metal atom, carbon dioxide requires a low-valent metal atom that is bound to electron-donating ligands that are large and bulky in the coordination sphere (*196*). When CO_2 does bond in the side-bound mode (Structure **3.26a**), its characteristics are similar to carbon disulfide bonding (*126*). Asymmetric and symmetric stretching frequencies for bound CO_2 generally fall in the ranges of 1500–1700 cm^{-1} and 1200–1400 cm^{-1}, respectively (*201*). Characterization of the bonding modes with IR spectroscopy is generally difficult because the CO_2 ligand

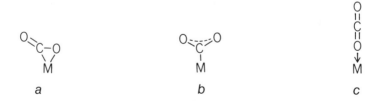

Structures **3.26a–3.26c**. Bonding modes of carbon dioxide.

can easily react to form carbonate (CO_2^-), bicarbonate (HCO_3^-), or carboxylato (RCO_2^-) species (*202*).

The nature of the coordination bonding of carbon dioxide was investigated in a series of frontier molecular orbital calculations. In these calculations, the three possible bonding modes of the ligand were compared in both a zero-valent nickel complex and a positively charged copper complex containing identical ligands (*203*). The calculations were designed so that the energy contributions of the individual bonding interactions such as back and forward donation, electrostatic effects and the distortion energy of the CO_2 ligand could be discerned. In general, for the zero valent complex, the side-on mode was the most stable. It derived its stability from both back bonding into the π^* level of the $C = O$ double bond and a ligand–metal electrostatic interaction. The distortion of the CO_2 ligand from linearity in the side-bonded mode added to the overall stability of the system because as the distortion increased, the π^* level dropped in energy and enhanced the back bonding interaction. This stabilization effect was limited, however, by the destabilization caused by the distortion itself. The opposing forces balanced at an O–C–O bond angle calculated to be 138°. The σ donation to the metal was not found to have a significant stabilizing effect (*203*).

In the positively charged complex, the end-on mode is favored because it is found to be stabilized predominantly by electrostatic effects (*203*). Back donation in the charged system could not be significant because the plus charge stabilizes the d orbitals and diminishes their ability to interact with the π^* levels of the CO_2 ligand. Bonding through the carbon atom is unfavored for both the zero-valent and positively charged species because the carbon atom cannot back bond. Back bonding is not possible because of the poor overlap of d and π^* levels in the zero-valent metal complex and because the carbon atom has a partial positive charge leading to a net repulsion with the cationic metal complex. The C-bound mode is thought to be stable, however, for the special case in which the metal center has little positive charge, but it is also a good π donor and is sterically crowded in a way that disfavors the side-on mode (*203*).

Another theoretical calculation (*204*) investigated the mechanism of formation of the side-bonded mode of CS_2 and CO_2 by determining whether the end-on mode (Structure **3.26b** and Structure **3.26c**) or the carbon atom bound mode (Structure **3.26b**) was the more favorable precursor. Because it is possible on surfaces to freeze out precursor bonding modes at low temperature, there is a possible link between chemisorption work and calculations of this nature.

These calculations were performed by first calculating the energy of a given precursor bonding mode and then, through a judicious variation of its bond angles, moving the $X = C = X$ ligand into the final side-bound mode. The energies of the intermediate states were monitored so as to determine the lowest energy pathway. The most likely pathway was found to be through the end-on configuration. Shifting the ligand off its axis required significant energy as well as bending the linear species toward the metal center. The stability of the ligand was regained, however, as the X–C–X angle decreased from 180°. The carbon-bound ligand was found to be much less stable than the end-on bound ligand, and because it also encounters a level crossing over the course of its transformation, it is not considered a viable precursor state (*204*).

On surfaces, carbon dioxide seems to bond as it does in complexes, although in some cases more strongly, and it apparently can undergo the scission reaction discussed for carbonyl sulfide. On clean Ag(110), TDS indicates that CO_2 does not adsorb at all. On oxygen-precovered Ag(110), CO_2 is very weakly bound and desorbs at 130 K (*205*). On dispersed platinum (*206*) and polycrystalline titanium (*207*) dissociation to CO and O_{ads} is detected. On rhodium dispersed on titanium dioxide or aluminum dioxide, CO_2 was found to form from CO and O_{ads}; in the presence of hydrogen, carbonyl hydride formed (*208*), a formation that was enhanced by impurities. On polycrystalline molybdenum (*209*) and W(100) (*210*), complete dissociation of CO_2 to adsorbed atoms was detected. Unlike the analogous reaction in metal complexes, the resulting O atom from CO_2 scission remained on the surface because it could not be removed by purification. CO_2 dissociation was thought to occur on Rh(111) (*211, 212*); however, that idea was subsequently attributed to the presence of a boron impurity in the metal surface (*213*). This result is consistent with the work on dispersed rhodium and a recent calculation (*214*) that found the dissociation probability per CO_2 collision on rhodium to be about 10^{-15} at low pressures. At high CO_2 pressures, CO_2 dissociation can occur significantly, even at such a low probability of reaction.

3.7.6   Sulfur Dioxide

In addition to the cumulenes that contain carbon as the central atom, sulfur dioxide, a molecule in which the central atom is sulfur, can also be classified as a cumulene ligand. The bonding in the free molecule differs from the carbon-centered cumulenes because the sulfur atom must use one of its $3d$ orbitals. This $3d$ orbital, along with its one $3s$ and three $3p$ orbitals, is necessary to accommodate the two additional valence electrons on the sulfur atom relative to carbon (Structure **3.27a**) (*37*). The two electrons fill a nonbonding orbital. This situation causes the triatomic molecule to be bent at a bond angle of 119.5° (*37*).

The differences in the electronic structure of the sulfur dioxide ligand can lead to a coordination chemistry in some cases different from the other cumulenes. Although sulfur dioxide coordinates to metal centers in the modes known for the other cumulenes, the side-bound mode (Structure **3.27b**), and the mode resulting in coordination through the central sulfur

Structures **3.27a-3.27f.** Structure of and bonding modes for sulfur dioxide.

atom (Structure **3.27c**), it also bonds in bridging modes that involve either the sulfur atom (Structure **3.27e**) or both oxygen atoms (Structure **3.27f**). More importantly, it bonds in a pyramidal mode (Structure **3.27d**), which is considered analogous to the bent diatomic structure of nitric oxide (*215*).

Similar to other ligands discussed, the coordination chemistry of sulfur dioxide is determined by the nature of its highest occupied and lowest unoccupied orbitals, which act as the electron-donor and electron-acceptor levels respectively. For SO_2, the HOMO is the $4a_1$ orbital (Figure 3.9, left) and the LUMO is the $2b_1$ orbital (Figure 3.9, right) which lies about 2 eV above the $4a_1$ level. The empty sulfur d orbitals do not participate in the bonding (*215*). For the sulfur-bound mode (also called monohapto planar), bonding occurs by donation from the $4a_1$ level and π-type back donation into the $2b_1$ level. The oxygen-bridged species bonds predominantly by π back donation into the $2b_1$ level; there is very little σ donation from the $4a_1$ level. The σ donor ligands enhance this interaction in metal complexes, whereas the π acceptor ligands diminish it. The pyramidal or bent SO_2

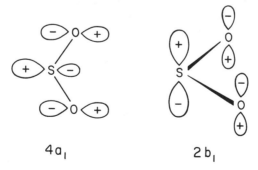

Figure 3.9. The HOMO ($4a_1$) and LUMO ($2b_1$) of SO_2, both of which can be involved in the coordination of SO_2 in a metal complex. (Reproduced with permission from reference 215. Copyright 1981.)

species bonds by a σ donation of electron density from the metal into the $2b_1$ LUMO of the SO_2 ligand, a situation analogous to bent NO (215).

The bent SO_2 structure is the most interesting of the sulfur dioxide bonding modes and is analogous to bent NO because of similar orbital overlaps and reactivity. The complexes containing this bent SO_2 species are the same d^8 square pyramidal species that often have a bent NO ligand at the apex of the pyramid. The complex shown in Structure **3.28** (216) is analogous to the bent NO complex (Structure **3.9a**). This finding indicates that the bonding requirements for bent SO_2 species are similar to those for bent diatomics. By extending the bending criteria for diatomics to SO_2, it may be surmised that the $4a_1$ acceptor level on the ligand has an energy that lies very close to the energy of the d_{z^2} level on the metal center that facilitates bending. Similar to NO and the diatomics, other donor ligands in the complex favor bending whereas other acceptor ligands disfavor bending (215). This finding indicates that, as for the diatomics, other ligands in the complex can affect the energy level of the d_{z^2} orbital and either enhance or diminish its interaction with the LUMO (the $2b_1$ level) of the SO_2 ligand. This situation is especially interesting because most sulfur dioxide complexes have this bent structure (215). Because of this large tendency toward bending, sulfur dioxide should be an excellent candidate for the study of possible bending processes on surfaces.

Structure **3.28**. Exemplary coordination complex illustrating the bent SO_2 coordination mode.

This facile bending ability of the SO_2 ligand gives it the ability to bond not only to metal centers, but also to other bound ligands (215). When SO_2 takes on the bent, or pyramidal, structure, it is acting as a Lewis acid, that is, an electron-pair acceptor. It can, therefore, bond, albeit weakly, to other ligands that have lone pairs available for donation; that is, Lewis bases. These ligands include species such as halides, I^-; azides, N_3^-; cyanates, OCN^-; oxo, O_2^-; and mercapto, $-SR$ (215). This ligand–ligand bonding interaction could lead to interesting possibilities regarding surface coadsorption and multilayer studies.

Like the other ligands discussed thus far, the various modes of sulfur dioxide coordination can be distinguished by vibrational spectroscopy. Although the asymmetric and symmetric stretching frequencies are lower, compared with the stretching frequencies for free SO_2, for all the bound modes of sulfur dioxide, the magnitudes of the frequency shifts are generally characteristic of a given bonding mode (*215*). The easiest structure to characterize is the side-bound mode (Structure **3.27b**) because the symmetric stretching frequency drops to a value below that for any other mode (between 850 and 950 cm^{-1}); this situation results in a splitting between the asymmetric and symmetric modes in excess of 290 cm^{-1}. For the sulfur-bound mode (Structure **3.27c**), the splitting is always less than 180 cm^{-1}, and the asymmetric and symmetric stretching frequencies lie in the ranges of 1300–1190 cm^{-1} and 1140–1045 cm^{-1}, respectively. For the bent SO$_2$ structure (Structure **3.27d**) the frequencies for the two stretching modes lie in the ranges of 1225–1115 cm^{-1} and 1065–990 cm^{-1}. The sulfur-bridging mode (Structure **3.27e**) has frequency ranges similar to those of the bent SO$_2$ structure, but it is much less labile than the bent structure. Thus, in this regard, the two modes can be distinguished. A similar situation exists for the vibrational spectra of the sulfur-bound and ligand-bound SO$_2$ modes. The ligand-bound species is reversibly bound and thus can be distinguished from the sulfur-bound mode. The oxygen-bridging species has been observed for only one complex, which has asymmetric and symmetric frequencies at 1043 and 919 cm^{-1}, respectively (*215*).

Few recent studies of SO$_2$ on surfaces exist. On Ag(110) (*217*), SO$_2$ chemisorption was studied by LEED, temperature-programmed reaction spectrometry (TPRS), and UPS. Adsorption at low exposures was molecular, and three binding states were detected. At intermediate coverages, the overlayer fell out of registry with the surface and SO$_2$ surface decomposition was detected. On Zn(0001) (*218*) and polycrystalline iron (*219*) SO$_2$ decomposed readily. SO$_2$ was formed on Pt(111) (*220*) by the reaction of oxygen with a (2 × 2) sulfur overlayer. No reports currently exist concerning the surface bonding modes of SO$_2$; however, considering the diversity of SO$_2$ bonding in complexes, such studies should be very interesting to pursue.

3.8   Heterolefins

Although ethylene and the cumulenes are olefinic species that bond to metal centers according to the Dewar–Chatt–Duncanson model, the heterolefin class of ligands is composed of doubly bonded species that usually bond through only the individual atoms that compose the heterolefin. Heterolefins include species with heterolefinic linkages such as those found in ketones and aldehydes (R$_2$C=O), dimethyl sulfoxide (R$_2$S=O), phosphites (R$_2$P=O), and imines (R$_2$C=N-R). The heterolefin ligand may also be considered an ambidentate ligand because it can bond through more than one atom in the ligand, depending upon the nature of the transition metal center.

3.8.1   Dimethyl Sulfoxide

One of the better understood ligands in this class is dimethyl sulfoxide (DMSO) (Structure **3.29a**), which has a pyramidal structure (*221*). The sulfur–oxygen bond length is 1.531 Å, 0.13 Å shorter than the sulfur–oxygen single bond, and implies the existence of some double-bond character (*221*). In metal complexes, dimethyl sulfoxide can bond either through the lone pair on the sulfur atom (Structure **3.29c**), or, more commonly, through one of the lone pairs on the oxygen atom (Structure **3.29b**) (*221*). For either bonding mode, the geometry of the molecule changes only slightly. The oxygen-bound species shows a metal–oxygen–sulfur bond angle of approximately 120°. This finding indicates that the lone pair is located in an sp^2 hybrid orbital on the oxygen atom (*221*).

a   b   c

Structures **3.29a-3.29c**. Structure of and coordination modes for dimethyl sulfoxide (DMSO).

Oxygen coordination is favored because the electron density of the π contribution of the S–O bond lies predominantly on the oxygen atom. This situation is shown in the valence bond pictures in Structures **3.30a-3.30c**, which show the three resonance structures for dimethyl sulfoxide. Structure **3.30a** shows the most favored valence bond structure in which the electron density of the π bond lies predominantly on the oxygen atom. In a valence bond sense, the S–O bond is composed of a sulfur atom that is sp^3 hybridized, in order to account for its tetrahedral geometry, and an oxygen atom that is sp^2 hybridized (Structure **3.30b**). The π bond arises by overlap of the filled, unhybridized, p orbital on the oxygen atom and an empty, low-lying d orbital on the sulfur atom. The magnitude of this overlap is heavily dependent on the electronic environments of the sulfur and oxygen atoms (*221*) because the bond involves the donation of the electron pair from the oxygen atom to the sulfur atom. Substituents can change the electron densities and therefore the charge on either of the two atoms. Thus, substituents can either increase or decrease the magnitude of the oxygen-to-sulfur electron donation. This effect can be detected by using vibrational spectroscopy because an enhanced oxygen-to-sulfur electron donation will increase the bond strength of the S=O bond and increase the bond's stretching frequency. A reduced electron donation will have the opposite effect.

Structures **3.30a-3.30c.** Valence band resonance structures for dimethyl sulfoxide.

Dimethyl sulfoxide is an interesting candidate for surface chemisorption studies (*222*) because this S–O peak shift in the vibrational data can be used to determine whether the molecule bonds through either the oxygen or the sulfur atom. The S=O stretching frequency is expected to decrease from its noncoordinated value of 1055 cm^{-1} when it is bound through the oxygen atom and to increase when it is bound through the sulfur atom (*221*). Oxygen coordination removes electron density from the oxygen atom. The resulting positive charge stabilizes the oxygen p orbital and thus diminishes the π bonding interaction between the oxygen p orbital and the higher lying d orbital on the sulfur atom. This situation results in a weaker S–O bond and a lower S–O stretching frequency. Sulfur coordination generates a partial positive charge on the sulfur atom, which stabilizes its empty d orbitals and improves the electron donation from the oxygen atom to the sulfur atom. These conditions lead to a higher S–O stretching frequency (*221*). Caution should be exercised when assigning the S–O stretch peak because the symmetric and asymmetric C–S–C stretching frequencies fall nearby at 680 cm^{-1}. Upon coordination, the C–S–C stretching frequencies also shift and can result in an incorrect assignment. This situation can be avoided by studying the hexadeutero ligand (*221*).

ESCA has also been used to determine the bonding mode of dimethyl sulfoxide. For an oxygen-bound moiety the splitting between the O 1s and the S 2$p^{3/2}$ is 365.8 eV, whereas for the sulfur-bound species, the splitting is 365.0 eV (*221, 223*). These splittings grow out of a set of electronic arguments similar to those employed to describe the peak shifts in the vibrational data for dimethyl sulfoxide.

It is, however, difficult to understand why S bonding ever occurs when the negatively charged oxygen atom, the superior donor, is so close. By becoming sp hybridized, it is possible for the oxygen atom to donate two lone electron pairs to the empty d orbitals on the sulfur atom and enhance the donor ability of the sulfur atom. However, the resulting structure is not thought to result in as stable a situation as direct donation from the oxygen atom to the metal. The most likely explanation results from the hard-soft acid-base (HSAB) theory. On the basis of the analysis of known dimethyl sulfoxide complexes, coordination through sulfur, a soft atom with large orbitals, is generally found for soft metal centers, whereas coordination through oxygen, a hard atom, is generally found for hard metal centers. Soft metals do occasionally bond to the oxygen atom, but probably because oxygen coordination results in a less sterically demanding geometry (*221*).

In addition to the two modes of coordination, dimethyl sulfoxide can also undergo S–O bond scission to the thioether, and, conversely, thioether can be oxidized to the sulfoxide in metal complexes. These reactions imply a side O-bonded ligand interaction, but as yet no structure of this type has been isolated.

Only one report exists concerning dimethyl sulfoxide chemisorption (*222*). By using both EELS and ESCA, it was found that dimethyl sulfoxide bonds through the sulfur atom on Pt(111). This finding is consistent with HSAB theory.

3.8.2   Phosphites and Phosphonates

A second ligand group in the heterolefin class is the phosphite ligand (Structure **3.31a**). Phosphites are similar in structure to dimethyl sulfoxide but have a phosphorus atom in place of the sulfur atom. The phosphites also have a hydrogen atom bound to the oxygen atom to achieve a balance of charge (*224*). Much like dimethyl sulfoxide, the phosphites can, after deprotonation, bond through either the phosphorus atom (Structure **3.31d**) or the oxygen atom (Structure **3.31c**). In addition, a phosphite bridging species in which bonding occurs through both the phosphorus and oxygen atoms (Structure **3.31e**) is known. Phosphorus coordination does not always require deprotonation. The phosphites are unique from dimethyl sulfoxide in that the R substituent can also be an alkoxy species as shown in Structure **3.31b** (*224*); these species are called phosphonates.

Structures **3.31a–3.31e**. Structures of and coordination modes for phosphite ligands.

The phosphite group of ligands have generated only limited interest possibly because of their limited availability (*225*). However, in relation to dimethyl sulfoxide, some comments can be made concerning phosphite coordination. First, the hydrogen atom is present (Structures **3.31a** and **3.31b**) because the phosphorus atom contains one less proton than sulfur, although the phosphite ligand contains the same number of electrons as dimethyl sulfoxide. The hydrogen atom appears to have no effect on coordination. When the species is deprotonated, a bond forms between a filled *p* orbital on the oxygen and an empty *d* orbital on the phosphorus atom, much as in dimethyl sulfoxide. This situation is supported by vibrational data of the P=O bond, which are similar to the data for dimethyl sulfoxide (*77*). The major difference between dimethyl sulfoxide and the phosphites is that coordination through the phosphorus atom is more common than coordination through the sulfur atom in dimethyl sulfoxide. This difference may be attributed to the fact that the additional plus charge in the nucleus of the sulfur atom, relative to the phosphorus atom, stabilizes the lone pair on the sulfur atom and makes the sulfur atom a poorer lone-pair donor relative to the phosphorus atom.

The characterization of phosphite ligands is done experimentally by ^{32}P-NMR because the vibrational peaks are broad and sometimes hidden by other more intense peaks. The P–O stretching frequency is the most characteristic for the phosphites; it falls near 1100 cm^{-1} for terminal P=O whereas the P–OH stretch falls near 880 cm^{-1}. The bridging P–O stretch falls near 1010 cm^{-1} (*224*).

It is difficult to comment on the feasibility of undertaking chemisorption experiments with phosphite and phosphonate ligands. Their commercial availability is limited (*226*), and the literature is silent concerning their volatility (*225*), although it is known that dimethyl methylphosphite [CH$_3$-P(O)·(OCH$_3$)$_2$] is a volatile liquid (*227*). The species with alkyl substituents are all probably volatile, especially the difluoro-substituted species. It may be possible to prepare these species on surfaces by the reaction of dialkylphosphines with preadsorbed oxygen. One study of dimethyl methylphosphonate on Ru(100) is known (*227*). The adsorption was dissociative to at least a small extent at 100 K, and complete decomposition of the adsorbate occurred when it was heated.

3.8.3 Oximes

A third heterolefin group is that of the oximes (Structure **3.32a**) (*228*), a species containing a C=N double bond with an –OH moiety attached to the nitrogen atom. The oximes differ from the dimethyl sulfoxide ligand in that no coordination to metal centers occurs through the atom bound to the substituent R group, that is, the carbon atom. Coordination does, however, occur through the nitrogen atom (Structure **3.32c**) in a manner similar to the oxygen coordination in dimethyl sulfoxide. Oximes can also bond through the oxygen atom (Structure **3.32d**) and bridging species (Structure **3.32e**) through the nitrogen and oxygen atoms are also known (*228*). However, little change in the C=N bond length occurs with coordination because, unlike dimethyl sulfoxide and the phosphites, the C–N π bond is composed of two half-filled p orbitals and is, therefore, not greatly affected by inductive interactions (*228*).

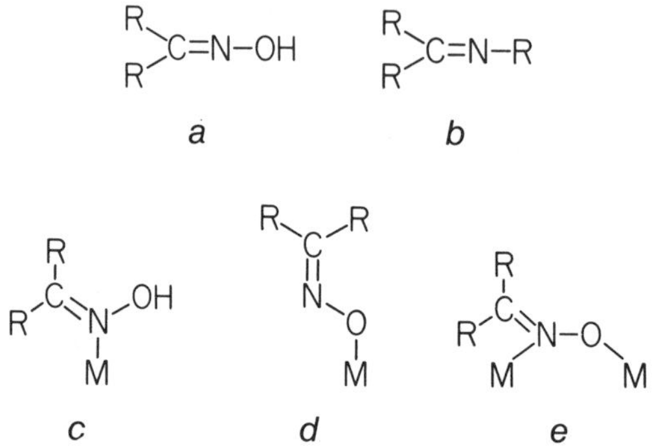

$$\underset{R}{\overset{R}{>}}C=N-OH \qquad \underset{R}{\overset{R}{>}}C=N-R$$

a   b

c   d   e

Structures **3.32a-3.32e**. Structures of and coordination modes for oxime ligands.

The imines are another type of species that contain C=N double bonds (Structure **3.32b**). They are similar to oximes, but the –OH group is replaced by an alkyl group (*54*). These species usually bond through the lone pair on nitrogen (*228*) and could also be considered members of the lone pair donors.

3.8.4   Formaldehyde and the Ketones

The one heterolefin ligand that has generated some interest in chemisorption studies is formaldehyde, $H_2C=O$. Species of this type are also known as ketones when the R groups are not hydrogen (Structure **3.33a**). Several examples of formaldehyde coordination can be found in the literature. In these systems the ligand either bonds through the oxygen atom (Structure **3.33b**) (*229*) or, unlike all the ligands in this class discussed so far, bonds in a side-on fashion through the C=O bond (Structure **3.33c**), much like ethylene (*229–233*). One well-characterized side-bound ketone is hexafluoroacetone (*234*), which is Structure **3.33d**. This species is unusually stable because the electron-withdrawing –CF_3 groups inductively stabilize the π^* level of the C=O moiety to an extent that leads to a back bonding interaction large enough to stabilize the ligand's coordination geometry. A few complexes do contain the unsubstituted side-bonded formaldehyde ligand, and some of their crystal structures are also known (*230–232*).

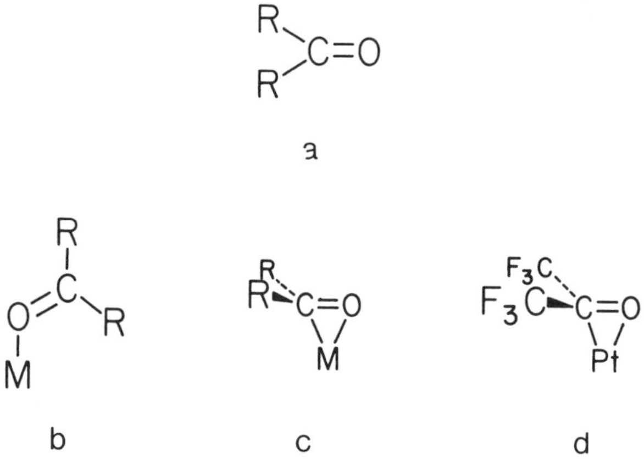

$$R_2C=O$$

a

b   c   d

Structures 3.33a-3.33d. Structure of and coordination modes for formaldehyde and the ketones.

On surfaces, formaldehyde and acetone species have been observed to bond through either the end on or side-bound modes at low temperatures, but they frequently decompose near room temperature. On polycrystalline tungsten (96), acetone bonds through an oxygen lone pair at low temperature. On Pt(111), EELS was used to determine that at low temperature, acetone bonds through the oxygen lone pair, whereas at higher temperatures (220 K), side-bonded acetone was detected (235, 236). On Ru(001) (235, 236), acetone was found to be side-on bound at 130 K. Because ruthenium can be easily oxidized, it is a better π donor and thus more able to stabilize a species with the π back bonding that occurs in side-bound structures. Hexafluoroacetone has been studied on Pt(111) (237) by using HREELS, and it was found that it bonds in a monohapto fashion through the oxygen atom. This finding indicates that, unlike the situation in metal complexes, the trifluoromethyl groups do not induce a π bonding interaction through the C=O π bond on surfaces (237).

3.9 Triatomic Pseudohalogens

The pseudohalogens are a class of polyatomic molecules whose properties are very similar to those of the halogens. They readily form dimers and strong acids with hydrogen and carry a –1 charge (38). The most common pseudohalogen is the diatomic cyanide anion (CN^-), which, for the purposes of this review, was grouped with the diatomic ligands. The triatomic pseudohalogens are grouped in a class because they are linear triatomics and have a unique coordination chemistry in comparison with the cyanide anion and the halogens. The pseudohalogens include species such as the azide anion (N_3^-), the cyanate anion (OCN^-), the thiocyanate anion

(SCN⁻), and the less common fulminate anion (CNO⁻), all of which are linear. All of the pseudohalogens bond in an end-on fashion, but, like the heteroolefins, thiocyanate is ambidentate and can bond in an end-on fashion through either end of the linear triatomic chain. All the pseudohalogens have been studied in detail by using vibrational spectroscopy (238), so only the fundamental trends in vibrational peak shifts will be discussed here.

Also included in this class is the linear triatomic nitrous oxide (N_2O), which, although it has only a meager coordination chemistry (239), has a potentially significant role as a surface adsorbate (240) because it is both isoelectronic with the triatomic pseudohalogens and is an electrically neutral gaseous ligand.

3.9.1 Azide

The azide anion has the valence bond structure shown in Structure **3.34a** in which the two outer nitrogen atoms are negatively charged and the center atom is positively charged. This situation results in a net charge of –1 (241). The valence bond picture of the azide anion and the resulting sp^2 hybridization of the outer nitrogen atoms are sufficient to provide a clear picture of the bonding of azide anions. Structures **3.34b** and **3.34c** show the two possible ground-state structures of hydrogen azide. In Structure **3.34b**, the H–N–N angle is close to 120° because of the interaction of the hydrogen atom with the sp^2 orbital on the outer nitrogen atom. In Structure **3.34c**, the H–N–N angle is closer to 109° and is caused by possible rehybridization of the nitrogen atom to sp^3 (241). However, linear H–N–N-bonding, which could result from the formation of sp hybrid orbitals on the nitrogen atom adjacent to the hydrogen is not possible because the central nitrogen atom is already sp hybridized and can accept no more π electron density.

$$^\ominus N{=}\overset{\oplus}{N}{=}N^\ominus \qquad \underset{H}{\overset{..}{\ddots}}\overset{\oplus}{N}{=}N{=}\overset{\ominus}{\overset{..}{N}} \longleftrightarrow \underset{H}{\overset{\ominus}{\overset{..}{N}}}{-}N{\equiv}N{:}^\oplus$$

a   b   c

Structures **3.34a–3.34c**. (a) Structure of the azide anion. (b) and (c) Resonance structures of hydrazoic acid.

The azide anion ligand can bond in three possible modes: It can be terminally bound (Structure **3.35a**); it can bridge bond through one nitrogen atom (Structure **3.35b**); or it can bridge bond through both outer nitrogen atoms (Structure **3.35c**). No side-on bonding through a π bond is known. In the terminal mode, the M–N–N bond angle falls between 117° and 132°. This finding indicates that the coordinated nitrogen atom is predominantly

sp^2 hybridized (*241*). When the angle is in excess of 120°, it is due to a certain amount of π electron donation from the nitrogen atom to the metal center. This situation is not possible for hydrogen azide because the hydrogen atom has no orbitals to accept the electron density. This π forward donation can be measured because, as the π bonding increases, the bond length of the N–N bond adjacent to the metal atom increases with respect to the outer N–N bond and thus leads to changes in the vibrational spectra. Of the two known bridging species that exist, the two nitrogen atom bridge (Structure **3.35c**) is more likely to form in complexes because bridging through one nitrogen atom can occur only when the metal centers lie close together (*241*). Unlike CO, the antibonding π^* orbitals of the azide ligand are not skewed to one end of the ligand. The reduced π^* orbital size at the coordinating end of the ligand requires, much as for dinitrogen (*115*), a closer approach to the metal centers to achieve an adequate back bonding overlap with the *d* orbitals (*241*).

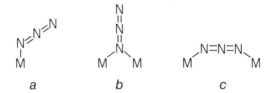

Structures **3.35a-3.35c**. Bonding modes of the azide anion.

The azide ligand is easily characterized by vibrational spectroscopy, and, because of the linear dependence of the ligand's π-forward bonding on the difference in the two N–N bond lengths, vibrational data can directly measure the degree of ligand-to-metal π bonding. The two predominant azide peaks observed are the asymmetric and symmetric stretches, which fall near 2000 and 1300 cm^{-1}, respectively. For the asymmetric stretch, as the N–N bond length difference increases, the stretching frequency increases; for the symmetric stretch, an increase in the N–N bond length difference leads to a frequency decrease (*241*). Although the symmetric stretch is rather weak in intensity, it is still observable by IR. A weakness of the vibrational data, however, is that they cannot distinguish the bonding mode of the ligand (*241*).

Because it coordinates readily in a bent fashion, azide is an interesting candidate for study on surfaces. It can be easily delivered to the surface by the hydrogen-capping procedure because hydrogen azide (hydrazoic acid) is quite stable as a gas and can be purified by distillation (*242*). On a surface, one of the two bridging modes may be favored over the terminal bent species, but careful selection of a surface could possibly disfavor both bridging modes. For example, the one-atom bridging mode (Structure

3.35b) may be disfavored by selecting a metal surface with a large inter-atomic spacing because the azide ligand, when one-atom bridge bonded, requires the metal atoms to lie close together. The two-atom bridging mode (Structure **3.35c**) could be disfavored by choosing a surface whose inter-atomic spacings are inappropriate for the two-atom bridge-bonding mode. If a terminal azide species could be found, EELS or IRAS could determine its degree of forward π bonding; ARUPS could also be applied to the system to study its degree of bending. It is also possible that the species would dissociate on a surface; experimental work would be necessary to resolve this issue. Azide has been reported as the product formed in a pulsed-laser-stimulated field desorption experiment of an overlayer of dinitrogen (*243*); this finding seems to be only the beginning, however, of an interesting area of chemisorption.

3.9.2   Thiocyanate

The most interesting of the pseudohalogens is thiocyanate (SCN^-) because it is ambidentate and is capable of bonding terminally through either the nitrogen end (Structure **3.36a**) or the sulfur end (Structure **3.36b**) of the molecule. This situation results in an N-bonded species linearly coordinated to a metal center and a sulfur-bonded species coordinated in a bent fashion similar to the azide ligand. Bridging thiocyanate species (Structure **3.36c**) are also known (*48*).

Structures **3.36a-3.36c**. Bonding modes of the thiocyanate anion.

As for the azides, the coordination mode of a thiocyanate ligand can be determined by using vibrational spectroscopy (*48*). The C–N and C–S stretches have frequencies of 2053 and 746 cm^{-1}, respectively, and are dependent upon the coordination mode. In addition, the bending angle for the S-coordinated species also affects the stretching frequencies (*48*). For nitrogen-bound thiocyanate, both the CN and the CS stretching frequencies increase upon coordination relative to the free anion; for sulfur-bound thiocyanate, on the other hand, the CN frequency remains unchanged with respect to the free anion, whereas the CS frequency increases relative to the free anion. The increase is, however, smaller than that in the nitrogen-bound case. With bending of the sulfur-bound ligand, the CN stretch frequency still remains unchanged, but the CS stretch decreases in frequency. For bridging thiocyanate species, the CN frequency is generally higher than that of the unbridged species. The frequency of the ligand bending vibrational mode is 480 cm^{-1} for the N-bound species and 420 cm^{-1} for the S-bound species (*48*). An alternate vibrational method for determining how thiocyanate is coordinated involves comparing the relative integrated intensity of the C–N stretching mode for the bound thiocyanate species to that of the unbound thiocyanate species (*244*). If the C–N stretching intensity of the coordinated thiocyanate ligand is less than that of the unbound thiocyanate species, then the ligand is probably bound through the sulfur; if the peak area is greater, it is probably bound through the nitrogen.

The major theoretical basis for the ambidentate behavior of thiocyanate ligands lies in the HSAB theory (*46*). As discussed earlier, similar orbital size is assumed to be a second order effect that enhances the bonding interaction between two orbitals. For thiocyanates, the geometrically large (soft) orbitals of the sulfur atom tend to bond with metal centers that also have large orbitals; the small (hard) orbitals on the nitrogen atom, conversely, favor bonding to hard metal atoms, which have relatively small orbitals (*46*).

There is also a Pd metal complex that contains one thiocyanate ligand bound through the nitrogen atom and one bound through the sulfur atom (*245, 246*). This compound is an exception to the HSAB theory, and this reflects the theoretical complexity of ambidentate bonding. The presence of both bonding modes in one complex can be explained through electronic effects because each thiocyanate ligand has a *trans* ligand with greatly differing back bonding abilities. The *trans* ligand in a complex competes with a given ligand for limited π back bonding. The N-bound thiocyanate occurs because its *trans* ligand is an excellent π back bonder. The *trans* ligand depletes the back bonding electron density at the thiocyanate coordination site and causes the poor back bonding N-bound mode to be favored. For the other thiocyanate ligand, the trans ligand is a poor π back bonder and leads to S coordination at the thiocyanate site, which is stabilized by significant back bonding.

The structural diversity of the thiocyanate ligand could lead to a very interesting surface chemistry. Like the azide ligand, thiocyanate is a stable gas when hydrogen capped (*175*), although it is somewhat difficult to prepare and purify. Chemisorbing H–SCN on surfaces composed of hard atoms may lead to a nitrogen-bound structure, whereas chemisorption on surfaces composed of soft atoms may lead to the sulfur-bound structure. The preference for two-atom bridging (Structure **3.36c**) may be less of a problem than it is for azide because one of the interactions on the surface will be either hard-to-soft or soft-to-hard and thus disfavor the bridge-bonding mode. Linkage isomerism of thiocyanate could also be investigated on surfaces; the possibility of carrying out the isomerization by ultraviolet (UV) radiation might be pursued.

3.9.3   Cyanate

The only pseudohalogen studied on surfaces so far is the cyanate anion (OCN^-) (*109, 247–249*). Although it is an analogue of thiocyanate, no oxygen-bonded species are known in metal complexes. The terminal nitrogen-bonded iso-species (Structure **3.37a**) is the most common (*106*), but the N-bridging isocyanate (Structure **3.37b**) and N–O-bridging isocyanate (Structure **3.37c**) are also known (*250*). The lack of oxygen-bound cyanate ligands has been attributed to the electronegativity difference between the nitrogen and oxygen ends of the ligand. Because nitrogen has the lower electronegativity, the orbitals on the nitrogen end of the ligand are larger than those on the oxygen end and thus allow a larger overlap with the metal orbitals (*250*). In spite of the lack of oxygen-bound ligands, calculations were carried out to predict the changes in the vibrational data, relative to the uncoordinated anion, for both the O-bound and N-bound cyanate bonding modes (*48, 251*). Although they are not as reliable as correlations from actual data, they provide a general picture for possible future needs. For nitrogen-bound cyanate, both the CN and CO stretching frequencies were found to increase upon coordination. However, for the theoretical oxygen-bound species, the CO stretching frequency increased, whereas the CN stretching frequency underwent no change. The CO stretching frequency increased for the nitrogen-bound case more than in the theoretical oxygen-bound case.

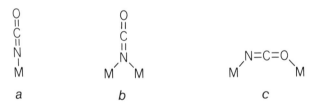

Structures **3.37a-3.37c**. Bonding modes of the cyanate anion.

The bonding of the cyanate ligand has been shown theoretically to be much like that of CO (*250*). Whereas CO has a 5σ donor orbital and 2π acceptor levels, the cyanate species has a 7σ donor orbital and 3π acceptor levels. The bonding differs from CO, however, in that the σ bonding interaction predominates in cyanate bonding, whereas the π-type interaction contributes very little to the bonding interaction. The energy of the e (3π) level in cyanate lies well above the back bonding metal d orbitals (the d_{xz} and d_{yz}) in the complex and thus causes a negligible bonding interaction.

Present surface investigations of the cyanate anion have been limited to preparing a surface cyanate species, and no attempts have yet been made to characterize its surface structure. Similar to thiocyanate, hydrogen capped cyanate is a stable gas (*252, 253*) and has already been used to adsorb cyanate on Rh(111) (*254*), Cu(111) (*247*), and unsupported chromium powder (*248*). On Rh(111), hydrogen-capped cyanate (cyanic acid) has been found by using TDS to have three adsorption states: a physisorbed state at 130 K, a chemisorbed state at 200 K, and a state in which decomposition to H_2, CO, N_2, and NH_3 occurs at higher temperatures (*254*). On clean Cu(111), cyanate did not adsorb at 300 K. On oxygen-precovered Cu(111), however, cyanate readily adsorbed and resulted in two ELS peaks at 10.4 and 13.5 eV. These peaks were attributed to surface cyanate because they did not resemble peaks for other surface species such as adsorbed oxygen, carbon, CN^-, NO, or N_2. The preadsorbed oxygen is thought to aid in dissociating the hydrogen ion from the cyanate anion (*247*). Surface-bound cyanate has also been produced from preadsorbed oxygen and both cyanogen (*109, 111*) and hydrogen cyanide (*249*) coadsorption. For these systems, the coadsorbed oxygen apparently reacts with one of the CN^- species. This reaction may be a possible preparation for the more interesting thiocyanate species because the coadsorption of sulfur and cyanide (HCN or cyanogen) is easier than the preparation of liquid thiocyanic acid (HSCN) (*175*).

In the organometallic literature, there is precedent for the preparation of coordinated isocyanate by reactions at the metal center. In one case, the addition of CO to a metal nitride yielded a cyanate ligand (*255*). Metal isocyanates have also been prepared by heating metal complexes in liquid urea $[(H_2N)_2C=O]$ at 130°C (*256*). At 130°C, the urea molecule decomposed, leaving the bound isocyanate ligand, ammonia, and water. Both of these methods may lead to the preparation of cyanate ligands on surfaces.

Methyl isocyanate, which has a methyl group bound to the coordinating nitrogen end of the isocyanate moiety, has been studied on both Pt(110) and Cu(110) (*257a, 257b*). It desorbed without decomposition on Cu(110); on Pt(110), however, 80% of the surface species decomposed. The molecule was found to chemisorb through the 2π a' orbital, leaving the methyl group unperturbed. When Cu(110) was predosed with potassium, decomposition of the methyl isocyanate to CO, NH_y, and CH_x species was detected for the sites near the adsorbed potassium atoms (*258*). A CO stretching frequency below 1600 cm^{-1} was taken as indicative of the formation of a salt with the surface potassium (*258*).

3.9.4 Nitrous Oxide

Although the pseudohalogens have an interesting coordination chemistry, they are difficult to handle and must undergo an as yet unproven hydrogen-capping procedure in order to chemisorb as pseudohalogens. Nitrous oxide (N_2O) (Structure **3.38**), however, is isostructural and isoelectronic with these pseudohalogens, and, because it is a gas, it can be much more easily used as an adsorbate in chemisorption studies and still provide information similar to the pseudohalogen data. Currently, only one complex is known in which nitrous oxide is a ligand (*239*), and although the bonding through the nitrogen atom is the favored structure, its geometry is still a controversial issue.

Structure 3.38. Valence bond structure of nitrous oxide.

$$^{\ominus} \ddot{N} = \overset{\oplus}{N} = \ddot{O}$$

Nitrous oxide is perhaps the only gas possessing a more extensive surface chemistry than coordination chemistry (*240, 259*). Recent surface studies of N_2O include a TDS, UPS, and XPS study on Ru(001) (*259*) and an EELS study on Pt(111) (*240*). On Ru(001), NO_2 dissociates at low coverage, but with increasing exposure below 180 K, it adsorbs associatively. He(II) photoemission indicates the absence of orbital bonding shifts with chemisorption, although the 7σ orbital intensity attenuates significantly (*259*). On Pt(111), N_2O was found to bond terminally through the nitrogen atom through a very weak interaction (*240*).

3.10   Organic Chelates

Up to this point, the ligands that have been discussed generally bond to one transition metal center through one electron pair. Some ambidentate ligands have also been discussed, but even though they can attach themselves to a metal center through more than one electron pair, they usually use only one of the pairs for coordination in a given bonding mode. The chelates, on the other hand, are a group of ligands that have more than one set of electron pairs that are oriented so that both sets can coordinate at the same time with one metal atom. The interaction requires a molecular species that has electron pairs oriented in such a way that each pair can occupy an individual coordination site on a metal atom and form a cyclic species containing the metal atom. The resulting cyclic species can have as few as three atoms or as many as seven (*242*); five-membered rings are the most stable. The number of chelate ligands used in organometallic chemistry is quite extensive, and because the nature of the bonding interaction is generally the same from ligand to ligand, only some of the more common organic chelating ligands will be discussed here.

Although chelates bond to metal centers through coordinate covalent bonds similar to those of nonchelating ligands, the bonding is generally stronger than two separate ligands bound to the same metal center by identical bonds. This added stability is due predominantly to the entropy increase when one chelating species replaces two nonchelating ligands in a complex. The effect can be visualized by realizing that when one of the ends of the coordinated chelate dissociates from the metal, it will most likely reattach itself to its original metal site because the other end of the chelate species keeps the dissociated end close to its original coordination site. This situation cannot occur for unchelated species (*242*).

3.10.1   Chelates Bound Through Lone-Pair Donors

One of the simplest chelates is hydrazine (Structure **3.39a**) (*77*), which forms three-atom chelate rings with transition metal centers although the rings are somewhat strained. One of the more common chelating ligands is ethylenediamine (Structure **3.39b**) (*77*), which forms the stable five-membered ring with a transition metal center. The superior stability of the five-membered ring chelate structure is due to the spatial distribution of the lone pairs in the ligand, which are optimally located for interacting with adjacent coordination sites of a metal center. A ligand similar to ethylenediamine is 1,4-diaza-1,3-butadiene (R–DAB) (Structure **3.39c**) (*260*). R–DAB illustrates a class of chelates that differ according to any number of substituent R-groups attached to the N=C–C=N skeleton. R-DAB can bond as a chelate through its lone pairs or through its C N π bonds in a Dewar–Chatt–Duncanson fashion (*55, 56*). These donor sites provide a total of four lone pairs for coordination, and structures are known where the ligand coordinates by using all possible combinations of some or all of these four electron pairs (*260*).

Structures **3.39a-3.39d**. Chelating ligands using the lone pairs on nitrogen and phosphorus atoms.

In the same way that ethylenediamine is analogous to monodentate ammonia, alkylaminobis(difluorophosphine) (AFP) (Structure **3.39d**) (*82*), is analogous to the monodentate phosphines. AFP is one of the few chelating ligands that have a substantial π acidity due to both the low-lying empty d orbitals on the phosphine and their added stabilization provided inductively by the fluorine atoms. In addition to its ability to chelate, AFP can also bridge between two adjacent metal atoms in a complex (82), as can many chelating ligands.

3.10.2   Negatively Charged Chelates

Unlike the amine- and phosphine-type chelates, many chelating ligands can carry a negative charge. An example of this type of ligand is the carboxylate ligand (261), which can bond to metal centers in not only the chelate structure (Structure **3.40a**), but also in a monodentate structure (Structure **3.40b**), in which only one of the electron pairs interacts with the metal, a monatomic bridge (Structure **3.40c**) and a diatomic bridge (Structure **3.40d**). The R substituent group can be almost any alkyl group or hydrogen; the bonding with the metal center, however, will usually not be dependent on the choice of substituent group. Vibrational spectroscopy can usually differentiate between monodentate and bidentate cooordination of carboxylates, but differentiation of chelate and diatomic bridging structures is not possible. For monodentate coordination, the C–O bonds are no longer equivalent, and the asymmetric and symmetric stretches (located at 1578 and 1414 cm^{-1}, respectively, for an uncoordinated carboxylate) increase and decrease their respective frequencies. Large splittings of these two levels in excess of the splitting found for the uncoordinated ligand, therefore, are usually indicative of monodentate coordination. On the other hand, chelate and bridging species show essentially no difference in their vibrational spectra relative to the uncoordinated ligand. For these cases, the symmetry of the ligand is unchanged, and thus only small frequency shifts are seen (261).

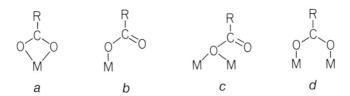

Structures **3.40a-3.40d**. Bonding modes of carboxylate ligands (formate, acetate, etc.).

An interesting carboxylate species is the trifluoroacetate anion (262), which can coordinate somewhat more strongly because of the electron-withdrawing effect of the trifluoromethyl substituent (Structure **3.41**). Vibrational characterization of this ligand is very difficult, however, because significant coupling between the CF$_3$ and CO$_2$ groups occurs, which leads to 15 separate modes, all of which can be detected by using IR spectroscopy

$$\underset{O^{\overset{\ominus}{\diagup}}\ddots O}{\overset{\overset{\displaystyle CF_3}{|}}{C}}$$

Structure **3.41**. The trifluoroacetate anion which can bond in the modes shown in Structures **3.30a-3.30c**.

(*262*). The oxalate ligand (*263*) is a special class of carboxylate in which the R substituent is replaced by another carboxylate species (Structure **3.42a**). Oxalato does not bond in the carboxylate chelation fashion; instead, it bonds through one oxygen atom on each individual carboxylate group (Structure **3.42b**). This bonding generates a more stable five membered ring chelate structure. If the coordination bond for the oxalate ligand is predominantly ionic, the symmetric and asymmetric stretches of the C-O bonds are between 1600 and 1650 cm^{-1}, respectively; covalently bound ligands have stretching modes between 1720 and 1750 cm^{-1}, respectively (*263*). In general, the coordinated C-O bonds lengthen (*263*); this lengthening results in a diminished C-O stretching frequency (*77*). For the unbound C=O bonds, an increased stretching frequency is usually detected (*77*).

a   *b*

Structures **3.42a** and **3.42b**. Structure and bonding mode of the oxalate anion.

3.10.3   Chelates Bound Through Sulfur

Carboxylate-type ligands that have sulfur atoms in place of the oxygen atoms are also known (*168b, 264*) and bond in the same fashion as the carboxylates (Structure **3.43a**). If the R group is an alkylamine (-NR$_2$), the ligand is called a carbamate (Structure **3.43b**), and if the R substituent is an alkoxy group (-OH), the ligand is called a xanthate (Structure **3.43c**). Another type of chelate with sulfur donors that is analogous to the oxalate anion is the dithiolate ligand (Structure **3.43d**) (*264*). Although the dithiolate ligand does not have two carboxylate type groups, it does have two sulfur atoms oriented so that a five-membered chelate ring can form upon coordination to a metal center. The R-substituent groups can be hydrogen, alkyl, trifluoromethyl (CF$_3$), or cyano (-C≡N), or can compose part of a benzene ring (Structure **3.43e**) (*264*). The simplest of the dithiolates is *cis*-1,2-ethylenedithiol (*265*) (where R = H, Structure **3.43d**), which could probably be used for chemisorption studies through hydrogen capping. Another disulfur ligand that could be studied on surfaces is bistrifluoromethyl ditiete (Structure **3.43f**) (*266*), a low-boiling liquid that may undergo S-S bond scission on a surface to form the dithiolate species. Dithiolates have three characteristic vibrational frequencies near 1400, 1110, and 860 cm^{-1}, representing the carbon–carbon stretch and the asymmetric and symmetric carbon–sulfur stretches, respectively (264).

Structures 3.43a-3.43f. Chelating ligands that bond through sulfur atoms.

3.10.4   Acetylacetone

Another common chelating ligand is acetylacetone (acac) (Structure 3.44a) (77, 267), a volatile solid that also has sulfur analogues (268). Acac can undergo a molecular rearrangement called keto–enol tautomerism in which the center carbon can lose a proton to the oxygen atom (Structure 3.44b). This loss results in a delocalization of negative charge through the five atoms of the resulting planar species. When it chelates to a metal center, the ligand bonds through the two oxygen atoms through one coordinate covalent bond and one σ bond (Structure 3.44c). The vibrational modes of acac are not characteristic of individual bonds, but stretching frequencies near 1570, 1530 and 930 cm^{-1}, attributable to C–C and C–O stretching, and 1450 cm^{-1}, are usually seen for oxygen-bound acac (77). A carbon-bound acac ligand is also known (Structure 3.44d), although it is less common than the oxygen-bound mode; it is characterized by stretching frequencies of the C–O bonds higher than those of the oxygen-bound species (77). Bonding through only one of the oxygen atoms is also known (77). The monothio acac species coordinate in the same manner as acac, although the vibrational data are more complex as a result of the lower symmetry of the species. In some cases, the metal–sulfur bond (2.26 Å) is shorter than the value calculated for it (2.49 Å) from the metal-to-oxygen bond distance added to the difference between the atomic radii of oxygen and sulfur. This result is the opposite of what is expected when atomic radii are considered. This situation has been attributed to the presence of significant metal d orbital to sulfur d orbital back bonding (268).

Structures **3.44a-3.44d**. (a and b) Illustration of keto-enol tautomerism of the acetylacetone (acac) ligand. (c and d) Bonding modes of the acetylacetone ligand.

3.10.5 Chelates on Surfaces

In general, most of the ligands discussed in this organic chelates section have received only a small amount of attention as chemisorption species on surfaces. Hydrazine was studied on clean and nitrogen-precovered Fe(111) by using UPS (*269a*). N-N bond scission was proposed to occur at room temperature. Dimethylhydrazine on Pt(111) was found to decompose by two pathways to either methane and nitrogen or to dehydrogenation products such as HCN or cyanogen (*269*). Ethylenediamine was found to decompose (Structure **3.45**) to cyanogen on Pt(111) (*69a*) through a nitrene intermediate, a structure in which each nitrogen atom loses its hydrogen atoms and bonds to two metal centers through σ bonds. Much of the work has centered around carboxylate species such as formate (*15, 270, 271*) and acetate (*64, 67*) species that generally bond in the symmetrical bidentate bridging mode (Structure **3.40d**). On oxygen-covered Pt(111), formate has been proposed to have the monodentate structure (Structure **3.40b**) (*16*). All of these surface studies used vibrational spectroscopy, but comparison of the data with that of the complexes is complicated by the surface selection rules, which often yield low intensities for the asymmetric OCO stretch mode in the C_{2v} symmetry at the carboxylate group common to surface carboxylates. The absence of an observable mode of this kind, however, is considered to be indicative of a symmetric C_{2v} surface species (*16, 271*). The symmetric stretching mode is observable, however, and usually falls in the region of the mode for transition metal complexes (*64, 270, 271*).

Structure **3.45**. Nitrene intermediate seen in the decomposition of ethylenediamine on Pt(111).

3.11　Nonmetal Oxyanions

The class of ligand that shows the greatest diversity of bonding modes is composed of the oxyanions of nonmetals. These include species such as NO_2^-, CO_3^{2-}, NO_3^-, and SO_4^{2-}, and usually carry negative charge. They can bond through either one or more of the oxygen atoms or through the central nonmetal atom, and bridging structures through any of the atoms in almost any combination are possible. Chelating structures as discussed earlier are also known for this class of ligand.

3.11.1　Nitrite

The oxyanion that best illustrates the wide variety of potential bonding modes of nonmetal oxyanions is the nitrite anion (NO_2^-) (*47*). In transition metal complexes, the nitrite anion has been found to coordinate in nine different ways, six of which are possible on surfaces. The nine include the N-bonded mode (Structure **3.46a**), two different terminal O-bonded modes (Structures **3.46b** and **3.46c**), a chelate mode (Structure **3.46d**), an N–O-bridged mode (Structure **3.46e**) and an O-bridged mode (Structure **3.46f**). In the remaining three structures, (Structures **3.46g-3.46i**), the nitrite ligand bridges between two metal atoms, again a situation with no apparent surface analogy. The nitrogen-bound species is called the nitro form, whereas

Structures **3.46a-3.46i**. Bonding modes of the nitrite anion.

oxygen-bound species are called nitrito forms. Surprisingly, no bidentate oxygen–oxygen bridge form has as yet been detected (*47*). This great diversity of structures is due to the presence of lone electron pairs on all three of the atoms in the ligand. This situation is shown in the two valence bond resonance structures of nitrite in Structures **3.47a** and **3.47b**. All of the lone pairs have an energy appropriate for donation to a metal center. In addition, the ligand can be either a π donor or a π acceptor, depending on how it is coordinated. When nitrite is coordinated through nitrogen, then the filled d orbitals on the metal can overlap with the empty π^* level of the N–O double bonds. When the ligand is coordinated through oxygen, however, the lone pairs not coordinated to the metal can donate electron density in a π fashion to the metal center (*47*).

a   b

Structures **3.47a** and **3.47b**. Valence bond resonance structures of the nitrite anion.

The reason nitrite bonds in these various modes can be rationalized in terms of the HSAB theory (*46*). For NO_2, the nitrogen atom is considered a soft atom, whereas the oxygen atom is considered a hard atom. When a given bonding mode forms in a complex, it can generally be described as a hard–hard or a soft–soft interaction. Much like thiocyanate, nitrite also forms linkage isomers, generally a nitro–nitrito set, in which one isomer is kinetically stable and slowly converts to the thermodynamically stable species (*47*). UV light has been found to reverse the process (*272, 273*).

The various bonding modes of the nitrite anion uniquely change the geometry of the ligand relative to the geometry of the uncoordinated anion. Nitrogen-bound nitro coordination shows little effect on the N–O bond length but increases the O–N–O bond angle. For a chelated species, the N–O bond length also does not change, but the O–N–O angle decreases. Monodentate nitrito coordination leaves the O–N–O bond angle untouched, but it increases the bond length of the N–O bond with the coordinated oxygen and decreases the uncoordinated N–O bond length. Bidentate nitrito coordination (Structure **3.46f**) causes a similar effect. For the N–O bridging mode (Structure **3.46e**), the one bridging N–O bond lengthens, whereas the other uncoordinated bond shortens, although the effect is not as pronounced as it is for nitrito coordination (*45*).

These structural differences between the different nitrite geometries are reflected in the vibrational data collected for each bonding mode. Nitrite has three normal modes, all of which are both IR and Raman active: the symmetric and asymmetric O–N–O stretches that occur at 1330 and 1260 cm^{-1}, respectively, and the O–N–O bending mode, which falls at 828 cm^{-1}. Nitrite is unusual in that the high interaction force constants cause the symmetric stretching frequency to lie at higher energy than the normally more energetic asymmetric mode (47). For the nitro structure, both stretching mode frequencies are higher than those of free nitrite, whereas the bending frequency is unchanged. For both the bridging and terminal oxygen-bound nitrito forms, the splitting between the two stretching mode frequencies increases relative to free nitrite, and the splitting of the monodentate bonding mode is the larger of the two. The bending mode is also unchanged. For the chelated nitrite bonding mode, the two stretching frequencies are unchanged relative to free NO$_2^-$, whereas the bending mode increases in frequency. For the N–O bridging structure, the splitting of the two stretching frequencies increases, and their mean values also increase; the bending frequency increases as well (47).

3.11.2   Nitrate and Carbonate

In addition to nitrite, a dioxide anion, several different types of trioxide species that coordinate to metal centers in similar ways are also known. These species include the sulfite dianion, SO$_3^{2-}$ (77), the carbonate dianion, CO$_3^{2-}$ (201) and the nitrate anion, NO$_3^-$ (274). The bonding modes of the nitrate anion (274) are common to the other species and include the unidentate mode (Structure **3.48a**), the chelate mode (Structure **3.48b**), and the bridging mode (Structure **3.48c**). The nitrate anion can also bond in a bent unidentate structure (Structure **3.48d**), which is unique to the nitrate species. In general, the unidentate species is common for the nitrates (274), whereas it is quite uncommon for the carbonates (275). For the sulfite dianion (SO$_3^{2-}$), coordination through the sulfur atom is also possible (77). In a manner similar to the carboxylate ligands, the N–O bonds that are coordinated through the oxygen lengthen, whereas the uncoordinated N–O bonds decrease in length relative to the uncoordinated species.

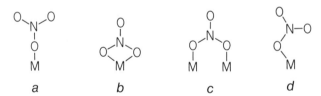

Structures **3.48a-3.48d**. Bonding modes of the nitrate anion.

Unlike for the other ligands discussed so far, IR spectroscopy is inadequate to differentiate between mono- and bidentate coordination because the two bonding modes have essentially the same vibrational spectra. Raman spectroscopy, however, can distinguish the two bonding modes because the polarization of the vibrations are unique to each structure (*274*). The mode that is used to distinguish the bonding modes is the doubly degenerate ν_3 mode of the free nitrate (or other) anion; it falls at 1390 cm^{-1} and represents the degenerate asymmetric stretching vibrations of the species. When the nitrate anion coordinates to a metal center, its symmetry is lowered from D_{3b} to C_{2v}. This change lifts the degeneracy of the ν_3 mode, which splits into an A$_1$ mode and a B$_1$ mode, of which the A$_1$ is polarized and the B$_1$ is depolarized. It has been found that for monodentate bonding, the B$_1$ mode is at the higher frequency, whereas for bidentate bonding, the A$_1$ mode is at the higher frequency. This difference can be easily detected by Raman spectroscopy (*274*). When studied on surfaces, the bonding mode of nitrate species could be determined in a similar manner; however, Raman spectroscopy would not be necessary because surface selection rules should discriminate between the allowed A$_1$ mode and the forbidden B$_1$ mode in the resulting C_{2v} surface complex (*276*).

As mentioned earlier, the carbonate anion (CO_3^{2-}) coordinates to metal centers in a series of modes similar to those of the nitrate anion. The only exception is the monodentate linear mode (Structure **3.48a**) (*201*), which is not found in any of the crystal structures determined for carbonate complexes. The carbonate ion does, however, have an extensive number of bonding modes in which the ligand bonds between two or more metal centers; Structures **3.49a-3.49c** illustrate a few examples of this type of bonding. When the carbonate anion is coordinated to a metal center, it, much like the nitrate anion, has a symmetry of either C_{2v} or C_s, which results in six IR and Raman active modes in its vibrational spectrum. These include the asymmetric and symmetric stretches (1536 and 1300 cm^{-1}, respectively), the C–O stretching modes (1069 and 1036 cm^{-1}), the CO_2 bending mode (in the vicinity of 773 cm^{-1}), and the nonplanar and planar rocking modes (near 857 and 773 cm^{-1}, respectively) (*197*). The splitting between the symmetric stretch and planar rocking modes is thought to arise from both ligand polarization and M–O bond strength; the splitting generally has been found to be greater for bidentate bonding (*201*).

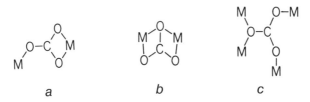

Structures **3.49a-3.49c**. Exemplary bridging modes of the carbonate dianion.

3.11.3　Sulfate

Oxyanions with four oxygen atoms can be represented by the sulfate dianion (SO_4^{2-}) which has both mono- and bidentate coordination modes much like those of the trioxide anions. They can be distinguished by using vibrational spectroscopy because monodentate coordination is associated with two vibrational bands whereas bidentate coordination is associated with three vibrational bands (277). The sulfate dianion ligand should be of interest to surface scientists because it can be bound in a complex, not only through reaction of the dianion with a metal center, but also by reaction of SO_2 gas with a dioxygen ligand already coordinated to a metal center (Scheme 3.4) (277, 278). The reaction can also occur by reaction of free oxygen with coordinated SO_2 in the pyramidal or bent bonding mode (215). This reaction has already been mimicked on a surface (73) and provides an easy method for preparing sulfate as well as sulfite species on surfaces. Similar reactions also occur between coordinated dioxygen and CO, NO, NO_2, PR_3, and alkyl isocyanides (277) and some of the products of these reactions have also been identified on surfaces (73, 205).

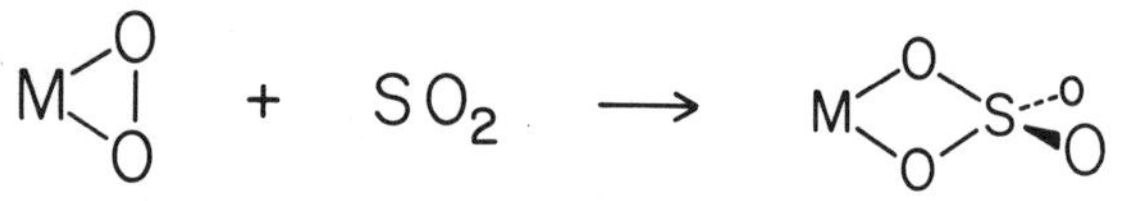

$$M\begin{smallmatrix}O\\|\\O\end{smallmatrix} \quad + \quad SO_2 \quad \longrightarrow \quad M\begin{smallmatrix}O\\ \\O\end{smallmatrix}S\begin{smallmatrix}O\\ \\O\end{smallmatrix}$$

Scheme 3.4. The preparation of coordinated sulfate by the reaction of coordinated dioxygen with sulfur dioxide.

3.11.4　Chemisorption of Oxyanions

Oxyanion species have received a fair degree of attention as chemisorbed species. NO_2 was studied on Pt(111) (279) and W(110) (280). While adsorption on W(110) led to a poorly defined overlayer, the Pt(111) surface caused a clean conversion to adsorbed NO and O. NO coadsorption with potassium on Pt(111), however, led to the reverse reaction because dissociating NO led to the availability of surface oxygen, which combined with surface NO to form NO_2 (281). On Pt crystallites dispersed on silica (282), low coverages of NO_2 yielded adsorbed NO and surface oxygen, whereas for higher coverages, surface nitrato species (NO_3^-) were also detected. When NO was added to preoxidized Pt crystallites, a nitrato bending mode was also detected (282). On preoxygenated silver powder (74), NO and NO_2 adsorption yielded NO_2^- and NO_3^- surface species. Reaction of SO_2 with oxygen-precovered silver powder (73) led to the formation of an SO_3^{2-} species in which all three oxygen atoms appeared to interact with the surface. A similar result was found on Ag(110) (217). Carbonate (CO_3^{2-}) was prepared on Ag(110) (205) by coadsorbing carbon dioxide with surface oxygen.

An additional adsorbate species that has a structure similar to these nonmetal oxides is nitromethane (H_3C-NO_2) gas, which has recently been studied on Ni(111) (*283*). It is believed to bond to the surface through both oxygen atoms in a bridging mode, much like the nitrate anion (Structure **3.50**).

$$
\begin{array}{c}
CH_3 \\
| \\
N \\
O \quad O \\
| \quad | \\
M \quad M
\end{array}
$$

Structure **3.50**. Bonding mode of nitromethane on Ni(111).

One potentially interesting area of study regarding the chemisorption of oxyanion species on surfaces is the effect of the resulting redox reactions of the chemisorbed species. For instance, when NO_2 is adsorbed on Pt(111), it dissociates into oxygen and nitric oxide (*279*); however, when NO is coadsorbed with oxygen on silver powder, NO_2^- is generated (*74*). A similar situation occurs for SO_2, which dissociates to sulfur and oxygen on polycrystalline iron (*219*) and forms from adsorbed sulfur and oxygen on Pt(111) (*220*). This difference may be due to the individual surfaces or to differences in the thermodynamic stability of the various product species. Such processes, though, apparently result from some redox processes on the surfaces and will have to be considered if further chemisorption work with oxyanions is to be undertaken.

3.12 Aromatics

The class of ligands with the most unusual type of bonding to metal centers is the aromatics (e.g., benzene). These cyclic species interact with a metal atom in a complex so that every atom in the cyclic system bonds to the metal center (Structure **3.51**). Although chemisorption studies with benzene are common, there are many other aromatic systems containing from 3 to 12 atoms that are, like benzene, aromatic and can quite easily coordinate to transition metal centers in the familiar flat-lying mode commonly known for benzene. These species offer a potentially wide range of chemisorption systems to be examined and may shed light on some new aspects of surface reactivity.

Structure **3.51**. Side-on bonding mode of benzene.

In the case of aromatic ligand coordination, the crucial point of understanding lies much more with the nature of aromaticity itself and how aromaticity occurs than with the actual overlap of d orbitals with ligand orbitals. This topic has already been discussed in great detail (*284*). The coordination chemistry of aromatics is now general enough so that it is understood that if a system is aromatic, it will coordinate to a metal center if appropriate conditions are found. Because a similar argument can be made for the surface bonding of aromatics, the basic concept of aromaticity will be explored in this section.

### 3.12.1	The Aromaticity Concept

Aromaticity can be defined as the stability gained through a cyclic system of conjugated double bonds, that is, a cyclic system of carbon–carbon bonds in which the double bonds alternate with single bonds (*185*). Aromaticity is most commonly found in cyclic species, but some noncyclics can also gain stability by the conjugation of double bonds. For a cyclic conjugated system a structure results in which each atom in the ring has one unhydridized p orbital. In a valence bond description, each p orbital pairs with another to form the conjugated chain; however, in the more accurate molecular orbital description, the entire set of p orbitals mixes together and forms a series of molecular orbitals that stabilize the molecule to a greater extent than would be obtained by the formation of localized single and double bonds. The added stability results because the p-orbital electrons (one from each p orbital) accumulate in the bonding molecular orbitals and thus maximize the π bond order. Cyclics with odd numbers of p-orbital electrons can gain or lose electrons as necessary in order to maximize their π bond order and thus attain aromaticity.

The existence of aromaticity can be predicted in a cyclic conjugated system through a molecular orbital picture of the cyclic system, which can be drawn qualitatively by following a set of general rules. First, the most stable of the molecular orbitals generated from the p orbitals is going to be nondegenerate. All remaining molecular orbitals will be doubly degenerate. If there is an even number of atoms in the cyclic, one orbital will remain, and it will be nondegenerate and have the highest energy relative to the other levels. The aromaticity of the conjugated cyclic species is then determined by the number of electrons that fill these orbitals. The general rule states that if the number of π electrons equals $4n + 2$, where n is an integer, then the conjugated cyclic species is aromatic (*185*). In other words, if the conjugated π system contains either 2, 6, 10, 14... electrons, then the cyclic species will show aromatic behavior. Generally, this statement means that aromaticity occurs only when the doubly degenerate bonding e states of the π system are filled; this situation leads to the four-electron interval between aromatic species. For benzene, the molecular picture will qualitatively be that shown in Figure 3.10 where the molecule's six π electrons, a number fitting the $4n + 2$ rule, fill the three lowest energy orbitals, the bonding orbitals of the π system.

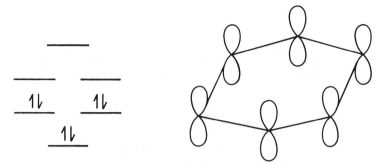

Figure 3.10. Molecular orbital diagram and *p*-orbital structure of benzene. Note the three filled bonding molecular orbitals.

However, suppose the cyclic molecule had only five atoms. Then its molecular orbital picture would be that shown in Figure 3.11 where only five *p*-orbital electrons fill the three π-bonding molecular orbitals formed from the *p* orbitals. If, however, this five-atom cyclic species were to add an additional electron, it would, like benzene, have six electrons in its π system and thus become aromatic. This situation has been observed to occur, and thus the cyclopentadienide anion ($C_5H_5^-$)(Structure **3.52a**) is aromatic and is known to coordinate to a metal center in ferrocene $[Fe(C_5H_5)_2]$ in the same way that benzene coordinates in dibenzene chromium (*285*). If the conjugated ring system had seven atoms, then the molecular orbital picture in Figure 3.12 would result. This cyclic species could also become aromatic, in this case, by losing the one lone electron and thus be left with only six π electrons, an electron count that will lead to aromaticity. This situation also is seen in the aromatic tropilium cation (Structure **3.52b**), which also coordinates to metal centers like benzene.

Figure 3.11. Molecular orbital diagram and *p*-orbital structure for cyclopentadiene. Note the half-filled bonding molecular orbital.

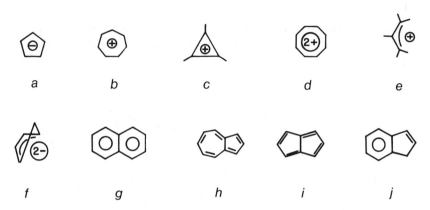

a   b   c   d   e

f   g   h   i   j

Structures **3.52a-3.52i**. Examples of aromatic hydrocarbon species that coordinate in a manner similar to that of benzene.

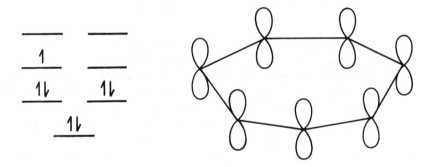

Figure 3.12. Molecular orbital diagram and p-orbital structure for cycloheptatriene. Note the extra electron in the antibonding orbital.

In addition to the molecular orbital picture, the qualitative shapes of the molecular orbitals that compose the diagram can also be determined. The most stable π orbital will have no nodal planes, and each e state to higher energy will have molecular orbitals with one additional nodal plane. For each e state, two possible structures can be drawn. The resulting orbital shapes for the benzene π system are shown in the left of Figure 3.13.

Extension of these arguments to a variety of conjugated cyclics results in a wide variety of aromatic species that can bond to metal centers as benzene does (*286*). Cyclics as small as the cyclopropenyl cation (Structure **3.52c**) and as large as the cyclooctatraene dication (Structure **3.52d**) are aromatic and thus bond in a manner analogous to benzene. It is even possible for noncyclic systems such as the allyl cation ($C_3H_5^+$) (Structure **3.52e**) (*287*) or even partially conjugated cyclic systems such as the π-dienyl type species (Structure **3.52f**) (*287*) to bond in the side-on fashion.

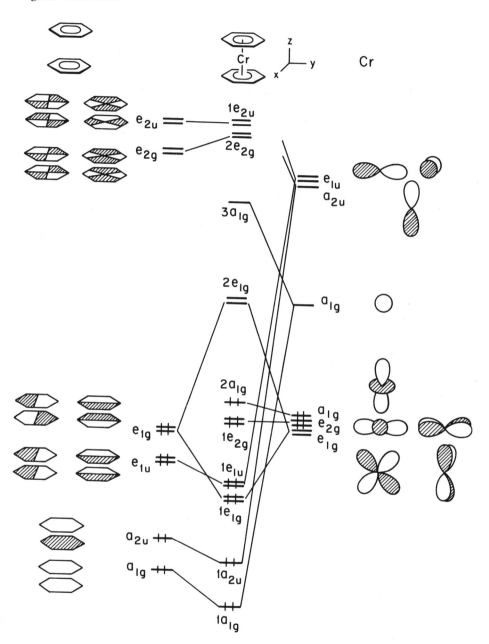

Figure 3.13. Molecular orbital diagram of the interaction of orbitals of the benzene ligands with the *d* orbitals of the chromium atom in chromocene.

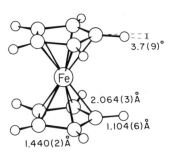

Figure 3.14. Illustration of the C–H out-of-plane bending of the cyclopentadienide anion. The C–H bond rotates toward the metal by about 5°.

In addition to monocyclic species, a wide variety of bicyclic systems such as naphthalene (Structure **3.52g**) (*286*), a molecule in which the π molecular orbitals span two fused cyclic rings, are also known for which each ring is able to bond to a separate metal center as benzene does. A similar bicyclic species is azulene (Structure **3.52h**), a species with a five-membered cyclopentadienyl ring fused to a seven-membered cyclohep-tatrienyl ring. Because the cyclopentadienyl ring must gain an electron to achieve aromaticity, and since the cycloheptatrienyl ring must lose an electron to become aromatic, a simple intramolecular electron transfer leads to two aromatic sites for metal coordination (*286*). Because the resulting separation of the two metal atoms coordinated to these two sites is the appropriate distance for a metal–metal bond, azulene is a very common ligand in dimetal systems. Other bicyclics include species such as pentalene (Structure **3.52i**), which is composed of two fused cyclopentadienyl rings, and indene (Structure **3.52j**), a structure containing a benzene ring fused to a cyclopentadienyl ring. Pentalene and indene, however, must add two and one electrons, respectively, to become delocalized for coordination to a metal center.

3.12.2 The Bonding of Aromatics in Metal Complexes

When a cyclic aromatic ligand coordinates to a metal center, it does so through its delocalized π molecular orbitals. Figure 3.15 (*284*) shows the specific molecular orbital interactions for dibenzene chromium (Structure **3.53**), a sandwich-type complex of D_{6h} symmetry. The left side of the figure shows the various molecular orbitals on the benzene ligands derived from the two sets of the six original *p* orbitals on the two benzene ligands. In the figure, although linear combinations are carried out for all 12 *p* orbitals of both benzene rings as a single unit, the specific orbital interactions are similar to the case in which, as on surfaces, only one benzene ligand bonds

Structure **3.53**. Structure of dibenzene chromium (chromocene).

to the metal. The right side shows the $3d$, $4s$, and $4p$ orbitals on the chromium atom split in the D_{6h} symmetry environment of the dibenzene chromium molecule. In Figure 3.13 it can be seen that the two sets of the three filled delocalized orbitals on benzene interact with the chromium orbitals as follows: First, the two lowest energy states of the two benzene rings, the a_{1g} and a_{2u} levels, donate electron density to the s and p_z orbitals of the chromium atom that have the corresponding symmetries. Second, the set of e_{1u} states on the benzene ligands donates electron density to the e_{1u} chromium levels, which are the empty p_x and the p_y orbitals. The third set of filled π orbitals on the benzene ligands, the e_{1g}, donates electron density to the e_{1g} symmetry d_{xz} and d_{yz} orbitals. According to this diagram, although each benzene ring has three bonds to the metal, none of them involve back donation from the chromium atom to the benzene rings (*284*). For ferrocene, the nature of the interaction of the cyclopentadienide anion with the iron atom is similar, even though the symmetry is D_{5h} (*286, 288*).

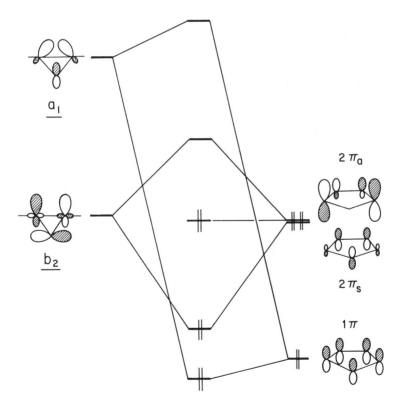

Figure 3.15. Molecular orbital picture of the bridging cyclopentadienide ligand. The structure of the metal site is to the left. (Reproduced with permission from reference 290. Copyright 1981 Academic Press.)

Coordination of benzene (or other aromatics) to a transition metal atom changes the benzene structure only slightly, but in a manner that may be significant to chemisorption work. The carbon–carbon bond lengths do not change appreciably; in free benzene, the carbon–carbon bond is 1.397 Å (*185*), whereas in dibenzene chromium, it is 1.398 Å (*284*). What does change is the rehybridization of the carbon atoms, which leads to a slight bending of the C–H bonds out of the plane of the ring. However, unlike ethylene, which undergoes a similar C–H bond bending away from the metal (*54*) when coordinated, aromatic species undergo a C–H bond bending toward the metal atom (*284, 285*). This situation is shown for cyclopentadiene in Figure 3.14. This bending is attributed to a hybridization effect that generates ligand *p* orbitals that point more in the direction of the metal atom than the unhybridized *p* orbitals. If this effect occurred on surfaces, it would probably lead to a metal–hydrogen interaction and subsequent dehydrogenation. Because benzene decomposition generally is a process that competes with desorption on surfaces (*19*), this rehybridization effect may also influence the bonding of benzene on surfaces.

A series of calculations were carried out for benzene chemisorption on Ni, Pt, and Ag (*289*), and it was found that for Ni and Pt, the out-of-plane bending should be 8° and 19°, respectively, away from the surface, whereas for Ag, the out-of-plane bending should be 2° toward the surface. These calculations also showed, however, that unlike its behavior in the metal complexes, benzene favors the threefold sites on surfaces. This finding may appear to be inconsistent with the organometallic chemistry–surface analogy. It is, however, probably due to subtle differences in the electronic structure of the metal center in compounds such as dibenzene chromium and those on a surface. This finding will be discussed in more detail in the conclusion of Chapter 4, in which the orbital structure of the two coordination sites is discussed.

3.12.3   Bridge Bonding of Aromatics

Cyclopentadienyl species can bond, not only to one metal center, but also to two metal centers in a bridging type structure (Structure **3.54a**)

a   b

Structures **3.54a** and **3.54b**. Illustration of (a) bridging cyclopentadienide and (b) the class of metal complex in which it appears.

(*290*) in complexes in which metal–metal bonding is apparently present. The vast majority of these ligand species are found in dipalladium complexes, but diplatinum complexes are also known. Unlike the monometal complexes, however, the ligand is only a four-electron donor and bonds to the metal atoms through three of the carbon atoms much like an allyl species does. This situation leads to a ligand rotation as the cyclopentadienyl ring changes the three adjacent carbon atoms that carry out the actual coordination bonding (*290*).

The bonding of the five-membered aromatic ring to the dimetal site occurs through donation from both the low energy, singly degenerate, a symmetry orbital (labeled 1π in the lower symmetry species) into the a_1 level of the dimetal fragment and the $2\pi a$ symmetry orbital into the b_2 level of the dimetal fragment (Figure 3.15). The dimetal levels also include an orbital contribution from the ligand opposite the aromatic ligand (Structure **3.54c**). Because electron density is lost from the $2\pi a$ level, an orbital that is bonding for two of the carbon–carbon bonds and antibonding for one, the bond lengths for the bonding C–C bonds should increase, whereas the bond length for the antibonding C–C bond should decrease. This situation is in fact the case (*290*).

In addition to dimetal bridging, aromatic ligands can also bridge three metal centers in triangular clusters, although only aromatic species with seven or eight atoms are large enough to accomplish this bridging. Five- and six-membered rings are usually limited to dimetal or edge-type bridging (*290*).

3.12.4 Characterization of Aromatics

The characterization of aromatic ligands is generally done either by crystallograph or by mass spectral analyses. Vibrational spectroscopies can do little to determine the nature or extent of aromatic coordination to metal centers because, unlike CO, there is no significant structural change in the molecule that is manifest because of the coordination interaction. Vibrational spectroscopy can, however, determine the nature of the bonding by determining the symmetry of the complex. The D_{5d} symmetry of a species like ferrocene (with a fivefold rotation axis) indicates rather conclusively that the bonding is that shown in Figure 3.14; the cyclopentadienide rings are side-on bonded to the metal center because no other geometry of the ligands can be of D_{5d} symmetry (*77*). Nuclear magnetic resonance (NMR), besides its use as a characterization tool, has been used relatively little to better understand the aromatic–metal interaction. The allyl ligand has been examined with ^{13}C-NMR (*291*) and it was found that the J(^{13}C, ^{13}C) coupling constant was the most sensitive to the nature of the metal–allyl bond. The J(^{13}C, ^{13}C) value was found to be directly proportional to the C–C bond force constant and inversely proportional to the C–C bond length.

3.12.5   Chemisorption of Aromatics

Chemisorption studies of aromatic species have, so far, been limited predominantly to benzene (*19, 62, 63, 101, 292–295*) and substituted benzenes (*101, 296a*). In every case, benzene has been found to lie parallel to the surface and has been observed to bond to 1 surface atom (*62*), 3 surface atoms (threefold hollow) (*19, 292*), and 8 or 12 surface atoms (*63*). On Ir(111), the surface complex of benzene has a C_{3v} symmetry with a trigonal distortion (*295*). Toluene chemisorption on nickel is much different from benzene chemisorption because chemisorbed toluene decomposes much more readily than chemisorbed benzene. This finding is attributed to the C–H bonds on the substituent methyl group that can more easily interact with the surface than the benzene ring C–H bonds and thus cause decomposition at a lower temperature (*101*). Azulene and naphthalene were studied by LEED and TDS on Pt(111) (*296b*), and it was found that several ordered phases exist as a function of coverage and that dehydrogenation occurs around 200 °C for both species. On Rh(111), naphthalene was found by LEED to have two flat-lying ordered structures separated by a temperature-induced phase transition. Dissociation occurred at 578 K (*294*). On the stepped platinum surface Pt(S)[7(111) × (100)], both azulene and naphthalene adsorbed molecularly and were flat lying because the terraces were wide enough to accommodate the bicyclic species (*297*). Two reports are available of a cyclopentadienide anion bound on Pt(111) (*18, 298*) above 250 K that has a flat-lying coordination mode similar to that found in ferrocene. The cyclopentadienyl ligand was placed on the surface by a partial hydrogen-stripping procedure because cyclopentene (C_5H_8) was the chemisorption gas that, upon chemisorption to the surface, dehydrogenated to cyclopentadienide ($C_5H_5^-$). Although attempts were made to use cyclopentadiene (C_5H_6) as the chemisorption gas, little success was achieved because of the apparent high acidity of the dissociated hydrogen atom (*299*). Hydrogen stripping or hydrogen capping is probably the best method for delivering to the surface charged aromatic species.

3.12.6   Inorganic Analogues of Aromatics

The cyclic species discussed so far contain only carbon atoms in their rings. A few ring systems, however, have no carbon atoms and thus are inorganic analogues of cyclic hydrocarbons. Although their bonding is not exactly like that of the aromatics, they are interesting species to compare to aromatic ligands. The most well-known of these inorganic analogues is borazine (Structure **3.55a**), the inorganic analogue of benzene. Although the species is nonplanar, the B–N bond lengths are smaller than those of ordinary B–N σ bonds. This finding indicates that some π-bond character is present in the ring. Apparently, this π-bond character results from the three filled lone-pair orbitals on the nitrogen atoms taking on some unhybridized

Structures **3.55a-3.55c.** (a) Structure of borazine, (b) analogous structure of a cyclic triazene, and (c) structure of pentamethylcyclopentaarsine.

p-orbital character in order to form a π system with the empty *p* orbitals on the boron atoms.

When borazine coordinates to a metal center, its bonding is more accurately described as three localized lone pairs that coordinate individually to unique coordination sites (*300*). The borazine ligand then becomes nonplanar and assumes the chair form. The B–N bond order also drops to that of a single bond. This occurrence indicates that coordination to a metal center eliminates the π character in the ring of the uncoordinated ligand. In addition to the benzenelike coordination, borazine also bonds to metal centers through σ-type bonding (discussed in general later) through either the boron or the nitrogen atom. M–B bonding generally occurs when the metal has filled *d* orbitals. M–N bonding generally occurs when the metal *d* orbitals are empty. This condition allows either metal-to-boron π back donation or nitrogen-to-metal *p* forward donation to occur and increases the stability of the metal-ligand bond (*300*). Another inorganic cyclic is pentamethylcyclopentaarsine (Structure **3.55c**) and the analogous pentamethylcyclopentaphosphine (*301*). The bonding of these species is analogous to that of borazine; it involves a nonplanar conformation and localized donation of three lone pairs to the metal center of the complex.

On surfaces, borazine should be a very interesting chemisorbed species because it has several possible modes of bonding. Because it can coordinate in the chair form, each B–N bond in the ring would, on a surface, have a component normal to the surface and thus could be detected more easily with surface vibrational spectroscopies. Species such as cyclic triazenes (Structure **3.55b**), could be studied in a similar fashion.

3.13   Heterocycles

Aromaticity occurs not only in conjugated cyclics of carbon atoms, but also in cyclics that contain atoms that are classified as nonmetals. These species are known as heterocycles. This type of ligand is unique because it can bond through either a lone pair on the heteroatom (in a manner like the lone-pair donor species discussed in the section on lone-pair donor ligands) with back donation if possible or, less commonly, through its aromatic π cloud, like the species in the class of aromatic ligands. The most well-known member of this ligand class is pyridine (Structure **3.56a**), which contains both a nitrogen lone pair and an aromatic π system for coordination (*47*).

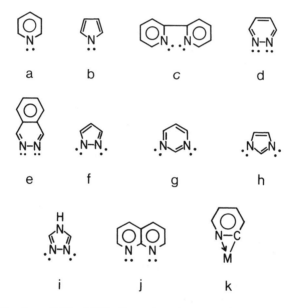

Structures **3.56a–3.56k**. Examples of nitrogen heterocycles.

3.13.1   Heterocycles of Nitrogen

The heterocycles containing nitrogen atoms include many species besides pyridine. Pyrrole (Structure **3.56b**) is the five-membered analogue of a pyridine ring, and 2,2'-bipyridine (Structure **3.56c**) (*77*) is a dimer of two pyridine molecules that acts as a chelating species. Heterocycles can also contain more than one heteroatom, as is the case for pyridazine (Structure **3.56d**), phthalazine (Structure **3.56e**), the pyrazolate ion (Structure **3.56f**), pyrimidine (Structure **3.56g**), the imidazolate ion (Structure **3.56h**), 1,2,4-triazene (Structure **3.56i**) and 1,8-naphthyridine (Structure **3.56j**), a heterocyclic species composed of two fused pyridine rings (*302*). For most of these, π coordination is rare, but through simultaneous coordination of their lone pairs, many unusual coordination complexes can be prepared. In addition, it is not uncommon for a hydrogen atom adjacent to a nitrogen atom to be lost. This loss results in a σ bond between a carbon atom of the heterocycle and the metal center (Structure **3.56k**) (*49*).

3.13.2   Heterocycles of Phosphorus

In addition to nitrogen heterocycles, there is a series of analogous phosphorus heterocycles (*303*) that have ring sizes from three to seven members. Examples of these species include phosphasine (Structure **3.57a**), the similar unconjugated phosphorene ligand (Structure **3.57b**), and phosphorin (Structure **3.57c**) all of which bond either through the π orbitals (Structure **3.57d**) that have significant aromatic character or the lone pair on the phosphorus atom (Structure **3.57e**) (*303*). The phosphorus heterocycles differ from their nitrogen analogues in that the phosphorus atom is never incorporated into a double bond. This situation leads to the presence of the extra nonring hydrogen or alkyl group bound to the phosphorus atom and results in a bent coordination mode in which the ring does not lie in the plane of the metal–phosphorus bond, as it does for pyridine (*303*).

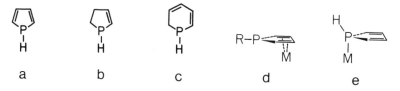

Structures 3.57a-3.57e. Examples of and bonding modes for phosphine heterocycles.

3.13.3   Heterocycles of Sulfur

Other heterocyclic ligands include species with sulfur heteroatoms such as thiophene (Structure **3.58a**). The sulfur atom on thiophene is a poor electron donor because of the aromaticity of the ring (*78*). Various multiheterocyclics of sulfur are also known such as isothiazole (Structure **3.58b**), which is an ambidentate species and bonds through either the nitrogen or the sulfur lone pair according to the HSAB theory (*46*).

Structures 3.58a and 3.58b. Examples of sulfur heterocycles.

3.13.4 Chemisorption of Heterocycles

Surface studies of heterocyclic species have been limited to studies of pyridine (*49, 295, 304–307*) and thiophene (*308–311a*) chemisorption. On Ag(111) (*305*), pyridine was found to be π bound at low coverage and underwent a compressional phase transformation to the weaker N-bound species that was tilted on the surface. On Pd(111) (*304*) a similar tilted structure was found with ARUPS. On Ir(111) (*295*), tilted pyridine was also characterized by using both ARUPS and ESDIAD. On Pt(111) (*311*), NEXAFS was used to characterize two different tilted phases of pyridine. At saturation coverage and low temperature, a tilt of 52° ± 6° was found, whereas at temperatures just below 300 K, a tilt of 74° ± 10° was found. Dimethyltetrazine was chemisorbed on Pt(111) (*312*), and two decomposition pathways, via either C–H bond scission or N_2 production, were characterized. On Cu(111) and Ni(111), thiophene was found to lie parallel to the surface (*308, 311a*).

The molecular species discussed in this section could be studied on surfaces in a manner similar to that of pyridine and thiophene. All the species discussed are either liquids or volatile solids and thus would present few problems with being delivered to the surface. Phosphorus heterocycles would be especially interesting to study on surfaces because their σ-bonding mode in complexes is so different from that of pyridine. It would be interesting to determine if the phosphorus-bound hydrogen is lost upon chemisorption through an interaction with the surface.

3.14 Electron-Deficient Species

Of the ligands that have been discussed so far, all have been shown to donate two electrons to a metal center and form a coordinate covalent bond. This donation results in the most common form of bonding in which two atoms are held together by a two-electron bond. A second form of bonding is known, however, where it is possible for three atoms to bond together with two electrons. This kind of bonding is called three-center two-electron bonding and is found in a class of molecules known as electron-deficient compounds, which includes some coordination compounds. Electron-deficient bonding is found most commonly in compounds that contain boron, an atom that has only three valence electrons. Three valence electrons are insufficient to provide an octet of electrons for the boron atom through conventional two-center two-electron bonding (*45*). When transition metal complex fragments bond to electron-deficient boron species, they can become involved in electron-deficient bonding and thus interact in a manner unique to that found in other coordination compounds. This situation raises a number of intriguing questions of how electron-deficient species will interact with surface metal atoms, a question as yet uninvestigated.

3.14.1   Hydride

The simplest ligand involved in electron-deficient bonding is the hydride anion (*313a, 313b*). It bonds in either the conventional two-electron two-center method when bound terminally (Structure **3.59a**) or in an electron-deficient manner when either bridge-bonded (Structure **3.59b**) or bound to a triangular face (Structure **3.59c**). Electron-deficient bonding in hydrides is most easily understood from a molecular orbital picture. For the case of bridge-bonded hydride, three orbitals, the filled *s* orbital of the hydrogen atom and one empty *p* orbital on each of the metal centers overlap and form three molecular orbitals (Figure 3.16). However, because only two electrons are present in the three-orbital system, only the lowest energy orbital of the three is filled. Because this orbital is a bonding one, bonding takes place at the three-center site (*313b*). A similar effect is expected when the hydride bonds to a triangular face, the only difference being the interaction of four orbitals instead of three, three of them being empty. Hydride ligands have been characterized principally by NMR spectroscopy (*314*). X-ray crystallography of hydride ligands is complicated by the poor detectability of hydrogen atoms, whereas neutron diffraction requires unusually large crystals (*313b*). Vibrational spectroscopy can characterize hydride ligand bonding to a certain extent; terminal M–H species have a stretching frequency between 1720 and 2160 cm^{-1} and a bending frequency between 650 and 850 cm^{-1} (*77*); however, in many cases, metal hydride modes exhibit low or undetectable intensities (*314*). Bridging hydrides in clusters can be indirectly located because they tend to increase the length

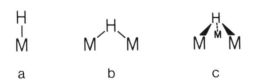

Structures **3.59a** and **3.59c**. Bonding modes of the hydride anion.

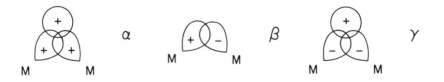

Figure 3.16. Molecular orbital overlaps illustrating the electron deficient two-electron three-center bonding of a bridging hydride ligand.   Only the α (bonding) orbital is filled.

of the metal–metal bond that they bridge (*313b*); however, this is not necessarily a reliable indicator.

The hydrides have been actively studied on surfaces for quite some time (*6*). On both W(100) and Ni(111), hydrogen atoms have been observed to occupy the hollow sites (*315*) where metal electron density is enhanced relative to atop sites. This bonding mode can be considered to be analogous to the electron-deficient bonding of hydrides found in transition metal clusters.

3.14.2   Boron Hydrides

The compounds that most commonly use three-center two-electron bonds in their structures are the boron hydrides. Unlike hydrocarbons, which tend to form chains, the boron hydrides form clusters in which the bonding is electron deficient (*316*). Transition metal ligand fragments interact with these clusters in one of several ways to form a variety of compounds. An example of a boron hydride species is pentaborane-9 (Structure **3.60**), which illustrates the three kinds of bonding found in boron hydrides. These include the terminal or exo B–H bond, which is a two-center two-electron bond, and both the bridging hydrogen and the cluster boron–boron bonding, both of which are electron deficient. The bridging hydrogen bonding is essentially the same as that for bridging metal hydrides; the boron cluster bonding is electron deficient because each B–H unit contributes three orbitals but only two electrons to the bonding in the cluster. This situation leads to the formation of a series of cluster molecular orbitals in which only the bonding orbitals are filled.

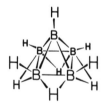

Structure **3.60**. Structure of pentaborane-9, which illustrates exo B–H, bridging hydrogen, and cage B–B bonds, the latter two of which are electron deficient.

When boron hydride species become ligands in transition metal complexes, it is possible for the metal fragment to bond in all of the ways discussed earlier. A metal complex fragment can replace a terminally bound hydrogen atom (Structure **3.61a**) or one of the bridging hydrogens (Structure **3.61b**). A metal fragment can also incorporate itself into the cage and, therefore, be a substitute for a B–H unit (Structure **3.61c**). Similarly, a metal center can also incorporate into the cluster by bonding to an open face. This occurrence eliminates the bridge-bonded hydrogens and increases the cluster size by one vertex (Structure **3.61d**). In addition to substituting itself for atoms within the cluster, a transition metal fragment can also bond to the bridging hydrogen atoms. This bond results in hydrogen bridges between the boron atoms of the cage and the transition metal in the complex atoms (Structure **3.61e**) (*316*). Different cage sizes tend to favor one or the other of these modes.

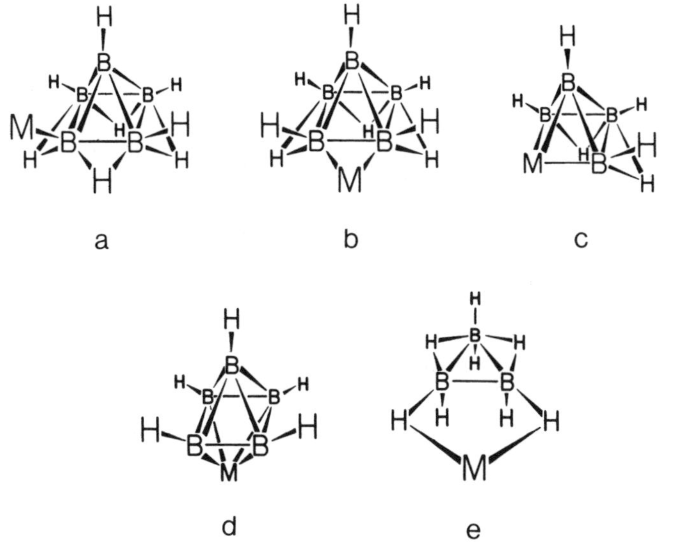

a   b   c

d   e

Structures **3.61a-3.61e**. Bonding modes of boron hydrides with metal centers.

A number of analogies may be drawn between the coordination structures just discussed and analogous hydrocarbon ligands. For instance, if diborane, B_2H_6, loses a proton, it becomes $B_2H_5^-$, which can bond to a transition metal fragment as shown in Structure **3.62a**. With the exception of the hydrogen atom that bridges the two boron atoms, the structure is analogous to side-on bound ethylene. For $B_2H_5^-$, however, there is less out-of-plane bending of the B–H bonds compared with the C–H bonds in ethylene, and the B–B bond length is greater than the C–C bond of coordinated ethylene. $B_2H_5^-$ also differs from coordinated ethylene in that the extra bridging hydrogen rehybridizes the boron $2p$ orbital so that it points more directly to the metal (*316*). This situation was similarly shown to occur for coordinated aromatics (*285*).

A species such as $B_5H_{10}^-$ (Structure **3.62b**) can also be considered a structural analogue of cyclopentadienide, again differing only in the presence of the bridging hydrogen atoms (*316*). Even clusters with more than five atoms can coordinate like cyclopentadienide if they have an open face with five atoms in a ring. The carborane $C_2B_9H_{11}^{2-}$ bonds in such a fashion, even though it is not isoelectronic with cyclopentadiene. Other a   b

Structures **3.62a** and **3.62b**. Boron hydrides that are analogues of hydrocarbons: (a) B_2H_5-coordinated in a manner analogous to ethylene and (b) $B_5H_{10}^-$, a species analogous to the cyclopentadienide anion.

carboranes (boron hydride species with carbon atoms incorporated into the boron cluster) include species such as $R_2C_2B_4H_5$ (*317*) and $C_2B_3H_7$ (*318*), species isoelectronic with cyclopentadiene. These species are also capable of bonding to a metal center in a manner analogous to aromatic species.

The nature of the interaction of boron hydrides with surfaces is essentially an open question because no reports have been found concerning boron hydride chemisorption. The boron hydrides are generally volatile, but the lower molecular weight species, the ones of greatest interest for chemisorption studies, are explosive in air and require special handling. When a boron hydride interacts with a surface it may react to form any of the ligand species found in transition metal complexes. If hydrogen capping is assumed to be operative, loss of either a terminal or a bridging hydrogen atom leads to the bonding modes shown in Structure **3.61a-3.61e**. It is also possible that if the square face of pentaborane-9 lost its four bridging hydrogens, a surface metal center may incorporate itself into the cluster cage. Another possible result could be the absence of any dissociation; lack of dissociation could lead to a situation involving a weakly bound surface species incorporating metal–hydrogen–boron bridging (Structure **3.61e**). If, however, the surface metal atoms do incorporate themselves into a boron hydride cluster, insight into the nature of the localized orbitals on the surface metal atom may be gained. For cluster bonding to form, the metal center must provide a certain number of orbitals and a certain number of electrons to the cluster (*316*). Whether or not a surface atom can meet these requirements, or if some metal surface atoms are able to do so whereas others are not may aid in elucidating the orbital structure of surface metal atoms.

3.15   Molecular Fragments as Ligands

Many of the ligands discussd so far in this chapter have been molecules that have a discrete existence when they are unattached to metal centers. There is also a group of species that bond to metal centers but are fragments of molecules and cannot exist unless bound to either metal centers or other atomic species. This class includes species such as methoxy (H_3C-O^-) and ethylidyne ($H_3C-C\equiv$), species that are fragments of methanol and ethane respectively. They are unique from other classes discussed in this chapter because although they can still be viewed formally as two-electron donors, they can also be viewed as species that are covalently bonded to the metal center. Species of this nature include fragments bonded through oxygen, nitrogen and most importantly, carbon. These fragment ligands have a significant propensity to π bond to the metal center in either a donor or acceptor sense, depending upon whether or not the p orbitals are filled.

3.15.1 Alkoxy Fragments

The alkoxy ligand ($R-O^-$) (319) and its sulfur analogue ($R-S^-$) (320) have been observed to bond to metal centers in either the terminal (Structure **3.63a**), doubly bridged (Structure **3.63b**), or the less common triply bridged (Structure **3.63c**) bonding modes (319). The bonding of alkoxy species is governed by the presence of three lone electron pairs on the oxygen atom, and the three bonding modes can be attained through donation of one, two, or three lone electron pairs to the metal atoms. However, the bond angles about the oxygen atom are appropriate to accommodate three lone electron pairs about the oxygen atom only for the threefold bonding mode when the bond angles are near 109° (319). The terminal and bridging modes show M–O–R bond angles of near 170° and 120°, respectively. These values indicate hybridizations that are less than the required sp^3, which indicates that filled unrehybridized p orbitals must be present on the oxygen atom for these two bonding modes. These filled p orbitals can then lead to substantial π donation of electron density from the electron-rich oxygen center to the metal in addition to the σ electron donation. For the bridge-bound species, the bond angle of 120° indicates sp^2 hybridization on the oxygen atom, a case in which one p orbital holds a pair of electrons and can donate them in a π bond to the metal. The 170° bond angle on the terminally bound species indicates the presence of almost two p orbitals available for π-type donation to the metal atom (319).

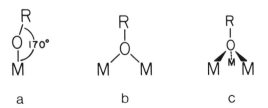

Structures **3.63a-3.63c**. Bonding modes of alkoxide species.

That the terminal alkoxy species has the most π bonding is corroborated by the observation that the M–O bond length is shortest for the terminally bound mode (319).

Although the bonding modes of alkoxy species are straightforward, characterizing them is unusually difficult. X-ray crystallography of alkoxy complexes is nearly impossible because single crystals of alkoxy compounds are difficult to prepare and are usually of poor quality. Vibrational spectroscopy is only somewhat better because alkoxy complexes tend to have low symmetries, a condition that makes band assignments difficult. In addition, the terminal and bridging modes have vibrational modes that lie very close in frequency. This situation further complicates structural characterizations (319). The carbon–oxygen stretching frequencies generally lie between 900 and 1150 cm^{-1}; for the case of the ethoxide ligand, the asymmetric stretch falls at 1070 cm^{-1}, whereas the symmetric stretch lies at 1025 cm^{-1} (319).

Vibrational modes attributable to terminal and bridging moieties have been assigned by reacting pyridine with alkoxy complexes that contain both bonding modes. It is believed that addition of pyridine converts all bridging alkoxy species into terminally bound moieties by blocking adjacent metal sites to bridging. For the case of $Nb_2(OCH_2CH_3)_{10}$, eight vibrational modes are detected at 1143, 1110, 1066, 1030, 914, 880, 575, and 485 cm^{-1}. Upon addition of pyridine, the modes at 1030, 880 and 485 cm^{-1} disappear; thus they are attributable to bridging ethoxide (319).

There have been several studies of alkoxide surface species because hydrogen capping provides an easy method for preparing the surface alkoxy moieties from alcohols. HREELS was used to detect methoxy (H_3C-O^-) on Pd(111) (321), Fe(110) (97a), and Ni(110) (66). On Cu(100) (322), IRAS was used to detect the methoxy species. The presence of both the symmetric and asymmetric C–H stretching frequencies was taken to indicate a bent surface methoxy moiety. On Ni(111) (323), however, ARUPS was used to show that the C–O bond in surface methoxy was nearly normal to the surface. On Cu(110) and Ag(110) (324), the ethoxy species was oxidized to formaldehyde, a process that was enhanced by preadsorbed oxygen. Studies of the surface reactivity of methoxy and ethoxy species on Ni(111) (68, 325) and clean and preoxidized Cu(111) (326) have revealed the various stages of bond breaking in the production and subsequent decomposition of these surface species.

The bent and linear methoxy groups found on Cu(100) and Ni(111) could be indicative of either different bonding modes or different magnitudes of π electron donation to the metal for the different systems. This question might be resolved by an accurate determination of the surface chemisorption sites through the pyridine coadsorption experiment discussed earlier. Metal–oxygen stretching frequencies are somewhat sensitive to bonding modes (319), but their frequencies are rather low (500 cm^{-1}).

A compound called pyridine 1-oxide (327) also bonds to metal centers through an oxygen–metal σ bond and an oxygen p to metal d π bond. Its bonding mode is shown in Structure **3.64** where the M–O–N bond angle is 120°, indicative of sp^2 hybridization of the oxygen atom. In single crystals of the complex, the pyridine ligand lies, as shown, in a plane almost perpendicular to the plane of the other ligands in the complex (*322*). This geometry minimizes the steric repulsion of the pyridine ring and the other ligands of the complex. However, the N atom of the pyridine moiety lies in the plane of the other ligands so that the *p* orbital of the oxygen atom can align itself for π bonding with the $6p_y$ orbital on the metal atom. This interaction, however, is probably not that strong and is unlikely to be maintained in the gas phase or in solution where no crystal packing effects are present (*327*). Pyridine 1-oxide may be an interesting species for chemisorption work because it is a volatile solid (mp = 65 °C) (*328a*) and coordinates in a bent fashion in complexes. It may also bond in this manner on surfaces. Also, it may be preparable on surfaces by the coadsorption of pyridine on an oxygen-precovered surface.

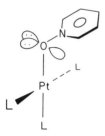

Structure **3.64**. Bonding mode of pyridine 1-oxide.

3.15.2 Dialkylamide Fragments

Species of coordinated nitrogen fragments, the dialkylamides, have two commonly known modes of bonding: the terminally bound mode (Structure **3.65a**) and the less common bridging mode (Structure **3.65b**) (*50*). The dialkylamide ligand is a fragment species because the nitrogen atom has only two alkyl groups bound to it. Neutral amines require three attached alkyl groups. The lone pair is used differently in the two bonding modes. It forms a $p\pi$–$d\pi$ bond in the terminally bound species and acts as a lone pair-donor in the bridging species (*50*).

The dialkylamides may be very easily prepared on surfaces through hydrogen capping with dialkylamines (R_2NH). This reaction has already

R⸝N⸜R
‖
M

a

R⸝N⸜R
⸝ ⸜
M   M

b

Structures **3.65a** and **3.65b**. Bonding modes of the dialkylamides.

been attempted on polycrystalline iron and nickel (*87*). Surface alkylamides, in which the R groups are hydrogen, have been prepared from ammonia on Ni(110) (*84*) and polycrystalline cobalt (*328b*). In both cases, the resulting species could only be approximated by an NH$_x$ species. ESDIAD has shown the species to be NH$_2$ on Ni(110); the plane of the NH$_2$ moiety lies perpendicular to the surface ridges. The species is believed to occupy the fourfold channel-spanning site (*85*). On Ni(111), N–H fragments have been found to form during the decomposition of hydrazine (*329a*). The N–H bond lies perpendicular to the surface and has a stretching frequency of 3340 cm^{-1} (*329a*). On W(110), NH$_3$, NH$_2$, NH, and N fragments have been found and appear to have unique N(1s) binding energies (*329b*). These values agree well with calculations (*329b*).

The use of alkyl-substituted amines may further aid in understanding amine chemisorption systems; nitrogen–carbon bonds probably will not undergo scission as readily as nitrogen–hydrogen bonds and may provide a method of predicting the structure of the surface species by assuming all N–H bonds are broken. Additionally, surface alkylamides may show an interesting surface atom dependence because transition metals with vacant d orbitals could provide a stronger π interaction with the filled p orbital on the nitrogen atom than transition metals with filled d orbitals. Surfaces such as copper that have nearly filled d orbitals may allow only a weak π interaction and result in unusual alkylamide structures.

3.15.3   Hydrocarbon Fragments

The most interesting of the ligand fragments are the metal–carbon σ-bonded hydrocarbon species. In transition metal complexes, hydrocarbon fragments have been observed to bond through a simple single bond (Structure **3.66a**) (*45, 65, 330–332*), bridge bonds between two metal centers (Structure **3.66b**) (*183*) or three metal centers (Structure **3.66c**) (*333*), a metal–carbon double bond (Structure **3.66d** and **3.66e**) (*334*), and, in a few cases, a metal–carbon triple bond (Structure **3.66f**) (*334*). A wide variety of bonding modes are also known that are composed of various combinations of metal–carbon σ bonds and Dewar–Chatt–Duncanson π-type bonding (*154*).

3.15.3.1   Hydrocarbon Ligands with Metal–Carbon Single Bonds   Metal–carbon single-bonded ligands are among the more difficult ligands to prepare, not because metal–carbon bonds are unusually weak as once believed, but because there are low-energy pathways leading to the decomposition of the metal–alkyl moiety. If these pathways can be blocked, then stable M–C bonding usually can be achieved (*330, 332*). The most common of the decomposition pathways is the β-hydride elimination mechanism illustrated

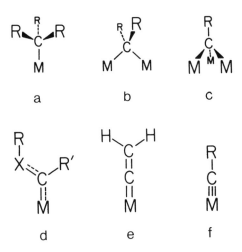

Structures **3.66a-3.66f**. Coordination modes of hydrocarbon fragments.

in Scheme 3.5. β hydrogens are hydrogens bonded to the carbon atom adjacent to the carbon atom that is bound to the metal center. The β hydrogens can interact with the metal center in the manner shown in the center of Scheme 3.5; this interaction results in a transfer of the β hydrogen to the metal and conversion of the alkyl group to an olefin, which may or may not remain coordinated to the metal atom. The pathway requires very little energy and occurs for almost any system containing β hydrogens (*330*). The mechanism can be blocked by using hydrocarbon species that contain no β hydrogens such as the neopentyl species (Structure **3.67a**); the methyl group (Structure **3.67b**) also has no β hydrogens and thus is a stable σ bonded ligand. Another mode of decomposition is reductive elimination (discussed later) in which two alkyl groups, σ bonded to a common metal center, could in a concerted fashion leave the metal center and bond together (*330*).

When decomposition pathways are taken into consideration, various methods can be used to prepare σ-bonded alkyl species in transition metal complexes (*45*). The simplest is the use of lithium or sodium salts of alkyl species. In these salts, the alkyl group has a minus charge and therefore a

Scheme 3.5.   Mechanism of β-hydride elimination.

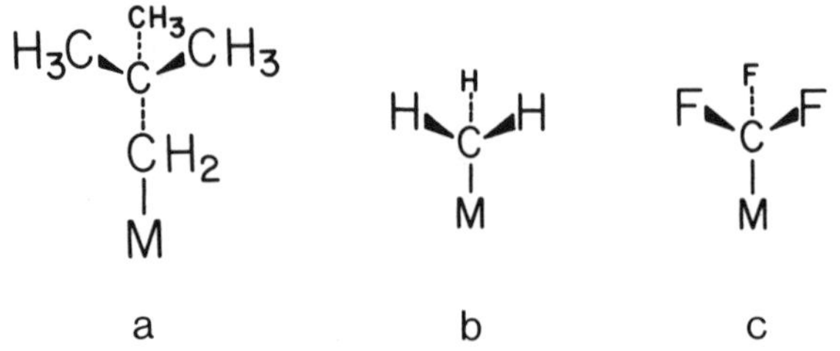

$$H_3C \underset{\underset{\underset{M}{|}}{\overset{|}{C}H_2}}{\overset{CH_3}{\overset{|}{\underset{}{C}}}} CH_3 \qquad H \underset{\underset{M}{|}}{\overset{H}{\underset{}{C}}} H \qquad F \underset{\underset{M}{|}}{\overset{F}{\underset{}{C}}} F$$

a   b   c

Structures 3.67a-3.67c. (a and b) σ-bonded neopentyl and methyl species; neither have β hydrogens and thus display unusual coordination stability. (c) σ-bonded trifluoromethyl species.

lone electron pair that can bond to the metal center in a manner analogous to the electron pairs in other charged ligands. The alkyl salts, however, are nonvolatile. In a similar reaction, the metal atom can replace an alkyl proton that is acidic. The alkyl species in this case also has a minus charge and an available lone electron pair for coordination. Metal alkyls such as those of mercury can also be used; these simply lead to an exchange of the mercury atom with the transition metal atom in the complex being studied. The direct reaction of an alkyl halide with bulk metal to yield metal alkyl halide complexes is another common method of preparing metal complexes with metal–carbon σ bonds. Few bulk transition metals can be reacted in this manner, but they are quite reactive with alkyl halides if the metal is prepared as an atomic species (45). This reaction, however, is less favored as the atomic number of the transition metal increases.

A more general route for the preparation of metal–carbon σ-bonded species that is similar to the reaction of alkyl halides with metals is oxidative addition, the reverse of which is reductive elimination (44). In an oxidative addition process, a given σ-type bond in a molecular species breaks and results in two molecular fragments. The fragments subsequently coordinate to a transition metal complex that has two coordination sites available for ligand bonding. The general reaction is shown in Scheme 3.6 where the ML_4 fragment is the coordinatively unsaturated complex and X–Y is the species that undergoes scission so that both X and Y can become ligands. The scission of X–Y bonds in the oxidative addition process can occur by several routes (44); the most interesting is the concerted cis-addition process in which the metal atom itself causes the X–Y bond to break. As

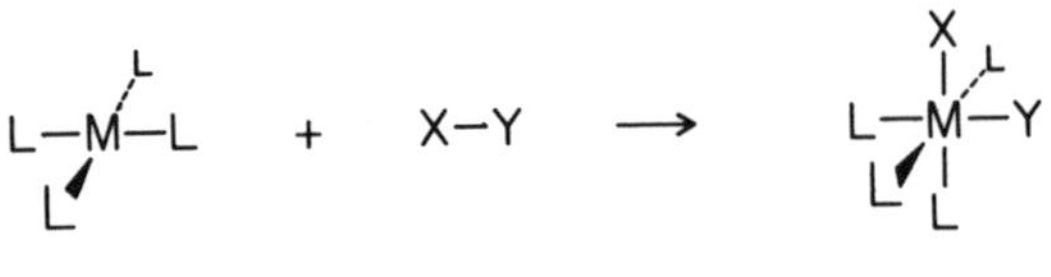

$$L-\underset{\underset{L}{|}}{\overset{L}{\underset{}{M}}}-L \quad + \quad X-Y \quad \longrightarrow \quad L-\underset{\underset{L}{|}}{\overset{X}{\underset{L}{M}}}-Y$$

Scheme 3.6. Illustration of the oxidative addition reaction.

shown in Figure 3.17, this process occurs through the donation of d electron density on the metal into the σ antibonding level of the X–Y bond (*44*). X–Y bond scission is favored because the X and Y fragments gain a bonding interaction with the metal as the X–Y bonding interaction decreases; the X and Y fragments are therefore never in the high-energy condition of being isolated radicals. Oxidative addition is favored, therefore, by high-energy metal complex species that have d orbital energies high enough to donate substantial electron density into the σ^* level of the diatomic. For this reason oxidative addition reactions are more likely for $5d$ metals than for $4d$ or $3d$ metals. Photolytically or thermally excited complexes may also be better suited to induce oxidative addition reactions (*65*).

Figure 3.17. Orbital overlaps illustrating possible mechanism of the oxidative addition reaction.

Species that undergo oxidative addition reactions include dihydrogen, H_2; hydrogen halides, HCl and HBr; alkyl halides (*44*) such as methyl chloride; and, most importantly, some C–H bonds (*65, 331*).

Of the methods available for preparing metal–carbon σ-bonded ligands in complexes, the most promising for the analogous reaction on surfaces is the oxidative addition reaction. It is the only reaction in which volatile species are involved, and thus allow a method for delivering the reacting species to the surface in vacuum. In addition, if the propensity for oxidative addition of C–H bonds can be exploited, a hydrogen-capping procedure results for the synthesis of surface metal–alkyl species. The other preparative routes generate unwanted byproducts that would remain behind on the surface and contaminate the resulting overlayer. This type of route includes reactions with the alkyl halides and the mercury alkyls.

Several examples of the preparation of surface alkyls that use routes analogous to those discussed for metal complexes have already appeared. Methyl chloride has been chemisorbed on Fe(100) (*335*) and XPS and TPRS were used to characterize the presence of surface-bound methyl groups. Although the chlorine remained on the surface, this experiment was a good indication of the potential of this surface preparative route. This preparative technique could be grouped with either the alkyl halide or oxidative addition reaction procedures. On polycrystalline Co and Ni (*336*), CH_3, CH_2, and C–H species were prepared by using a similar method. Reactions that fall under the group for the oxidative addition of C–H bonds include the chemisorption of cyclopropane (*337*), ethane, propane, isobutane, and neopentane (*338*) on the 110 surface of iridium. No specific surface structures have been characterized for these systems, but their complete decomposition to hydrogen and carbon at elevated temperature implies a

C–H scission process, promoted by the surface, that leads to a metal–alkyl intermediate species. In a recent study (*325*), surface methyl species were proposed as intermediates in the below room temperature decomposition of ethanol on Ni(111).

Oxidative addition reactions are quite common on surfaces. It is reasonable to propose that H_2 dissociation occurs on surfaces by the mechanism described in Scheme 3.6. In addition, the halogens, dioxygen, and dinitrogen species all dissociate on surfaces. These processes could also be reasonably described through an analogy of the concerted cis-addition process described in Scheme 3.6. From this perspective, it is reasonable to believe that metal–carbon σ-bonded species could be prepared on surfaces through oxidative addition.

A metal–carbon σ-bonded ligand with unique preparation characteristics is the trifluoromethyl ligand ($-CF_3$) (*131*), which bonds only in the terminal fashion (Structure **3.67c**). Although in metal complexes the $-CF_3$ ligand is known predominantly for nontransition metal species, it is prepared in a manner that may be easily applied to surface chemisorption studies. Trifluoromethyl radicals can be prepared by passing hexafluoroethane (F_3C-CF_3) through a radio frequency discharge of 25 W at a frequency of 8.6 MHz (*339*); the hexafluoroethane pressure is kept at 1 torr. The C–C bond in hexafluoroethane is weak relative to the C–C bond in ethane and is easily broken in the discharge. The resulting radicals are then immediately reacted with metal halides to form metal trifluoromethyl species. It is possible that similarly generated $-CF_3$ radicals could be trapped out on clean single-crystal surfaces in a vacuum. The presence of CH_3 on Fe(100) (*335*) is an ample precedent for the stability of surface methyl moieties and a very interesting surface chemistry may be uncovered.

Another interesting reaction that may have a surface analogy is the fluorination of methyl groups with a dilute fluorine gas (*131*). An example reaction is the following:

$$H_3C-Hg-CH_3 \xrightarrow[-110°C]{He/F_2} F_3C-Hg-CF_3$$

The gas contains a few percent of F_2 in helium, and the reaction occurs readily and must be slowed by cooling the reaction to -110 °C. On surfaces, this reaction could be quite favored, and it would be interesting to determine the applicability of the fluorination reaction with regard to various surface hydrocarbon moieties. Although the chemisorption of fluorine to the metal would also be expected, it may be retarded by the site blocking caused by the adsorbed hydrocarbon.

A class of molecular species that is an excellent candidate for preparing metal–carbon σ bonds on surfaces is composed of the ylides (*340, 341*). The ylides have the general structure shown in Structure **3.68a**, and they have a ylene resonance structure shown in Structure **3.68b**. The ylide structure has four alkyl groups bound to a central atom, which could be phosphorus, sulfur, or arsenic (*341*). One of the alkyl groups is a methylene group (CH_2) that carries a minus charge and an electron pair in an unhybridized *p*-orbital. This condition leads to a phosphorus–carbon multiple bond much like the $S=O$ bond found in dimethyl sulfoxide (page 64). The filled *p* orbital on the carbon atom donates its electron density into a low-lying empty *d* orbital on the phosphorus atom. There is a greater tendency, however, for the negatively charged carbon atom to bond to another positively charged center. When the ylide species bonds to a transition metal center they form a metal–carbon σ bond (*341*), much like the negatively charged alkyls of lithium. It is likely that ylides would react in a similar fashion on transition metal surfaces. Ylides are generally volatile (*342*), thus they would present no problems in the preparation of the surface overlayer.

a   b

Structures **3.68a-3.68b**. Valence bond pictures of (a) an ylide and (b) the corresponding ylene structure.

In general, ylides bond much like the chelates in either a terminal (Structure **3.69a**), bridged (Structure **3.69b**), or chelating (Structure **3.69c**), fashion. More than one carbon atom can coordinate to a metal center because as many as three of the methyl groups in the ylide can lose a proton and support a negative charge (*340*). Ylide complexes are generally quite stable and even form stable compounds with metals such as silver, copper, and gold (*340*).

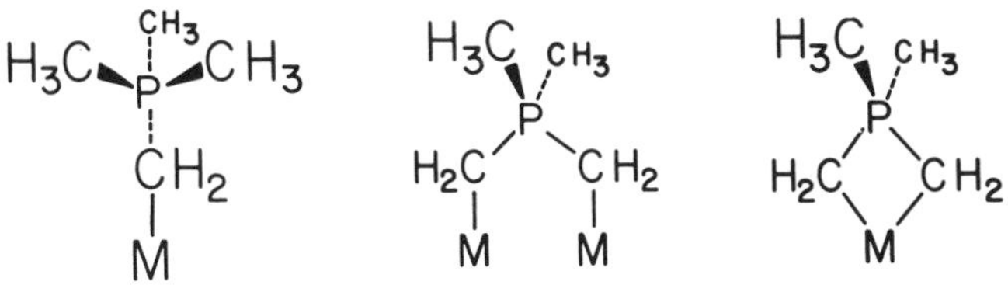

a   b   c

Structures **3.69a-3.69c**. Bonding modes of ylides.

3.15.3.2   Twofold Bridging Alkyl Fragments In addition to forming one metal–carbon σ bond, alkyl groups can also form two σ bonds to two metal centers in a bridging mode (Structure 3.66b). Although electron-deficient species are known in which a methyl group bridges two metal centers (*45*), the most common type of alkyl bridging involves a CH_2 or CR_2 bridge (*183*) in which, in a valence sense, the carbon atom gains its octet of electrons from two σ bonds to hydrogen atoms and two σ bonds to two individual metal atoms. These structures, known as methylene bridges, are generally symmetric and have M–C–M bond angles within the range from 70° to 81°. Such a bond angle is indicative of a strong metal–ligand interaction because the angle is much smaller than the expected tetrahedral angle near 109°. The R–C–R angle in the methylene-bridged moiety usually falls between 104° and 110°, the expected value for tetrahedral bonding (*183*).

Although the bonding of methylene ligands can be adequately described by this valence bond approach, molecular orbital theory offers a more realistic picture of the bonding, which is in part electron deficient (*183*). The methylene bridge, as is the case for many other ligands, has both a σ donor orbital of a_1 symmetry and a π acceptor orbital of b_1 symmetry. As shown in Figure 3.18a, the donor orbital is simply a filled sp^2 hybrid level, whereas the acceptor orbital (Figure 3.18b) is the empty unhybridized p orbital (*343*). The a_1 orbital forms a donor bond with an a_1 symmetry combination of d orbitals from the dimetal bridge site, whereas the back bonding occurs from the combination of the metal d_{z^2} orbitals of the dimetal site (of $1b_{3u}$ symmetry) into the b_1 orbital or the methylene bridge (*343*). The b_1 orbital of the methylene group has a slightly lower energy than the metal d orbitals, so the back bonding is quite extensive and is much more important to the bonding interaction than the electron-deficient donor bond (*183*). This result is consistent with the valence bond picture because valence bond theory requires the carbon atom to acquire two additional electrons from the metal to achieve its octet. The molecular orbital picture, however, is more descriptive of the bonding because the presence of back donation can be used to explain the variations of the metal–carbon bond length found in metal complexes containing methylene bridges. Donor and acceptor ligands on the metal center enhance and attentuate back bonding,

a   b

Figure 3.18. (a) σ donor orbitals and (b) π acceptor orbitals of the methylene ligand.

respectively, and shorten or lengthen the metal–carbon bond of the methylene bridge (*183*).

Methylene is both a better σ donor and π acceptor (*183*) than CO. This finding indicates that once methylene is bound to a metal bridge site, it is more strongly bound than the CO ligand. The rarity of methylene bonding is probably due, therefore, to greater difficulties regarding its preparation.

Numerous methods can be used to prepare transition metal–methylene bridge complexes (*183*). The methods include the use of dihaloalkanes, a reaction analogous to the methyl chloride reaction discussed earlier, except that two C–Cl bonds are broken instead of one. Indirect methods are also known, such as the interaction of metal centers with metal–carbon multiple bonds (discussed later) (*183*). The most important of the preparative routes involves the use of diazomethane (H_2CNN) (*183*), a volatile species in which the C–N bond can be easily cleaved. This method results in an essentially inert dinitrogen molecule and a methylene fragment ready to be coordinated. It is especially valuable because it is carried out at ambient temperatures (*183*). A molecular species that reacts in a manner similar to diazomethane is ketene ($R_2C=C=O$). Ketene has a methylene group attached to a $C=O$ species that can, upon cleavage of the methylene species, form carbon monoxide (*183*). In addition, dialkyl ketene can also be deoxygenated in the presence of CO to yield CO_2 and the bridging alkyl species (*200*).

So far, only four reports exist concerning the preparation of methylene bridges on surfaces. In one report (*184*), diazomethane was passed over various transition metals; the major products were nitrogen and ethylene, species that were formed presumably by the joining of two surface-bound nitrogen atoms or methylene species. The addition of hydrogen led to hydrocarbon chains ranging from 1 to 18 carbon atoms. On Pt(111), the chemisorption of diazomethane led to a series of decomposition products consistent with a surface methylene intermediate (*344a*). Diazomethane was also chemisorbed on Ru(001) (*72*), and because the resulting EELS spectrum compared favorably with that of the C–H modes of diiodomethane (methylene iodide, H_2CI_2), methylene was proposed as the surface hydrocarbon species. In a third report (*344b*), methylene was prepared on Fe(110) by exposing the surface to ketene. These three successes are encouraging for the use of diazomethane and ketene on a variety of surfaces. Use of these molecules is complicated, however, by their high reactivity with metal surfaces, which makes handling difficult. In the Ru(001) study, however, the metal gas handling systems were preconditioned with diazomethane until reactions with the walls became negligible (*72*).

3.15.3.3   Threefold Bridging Alkyl Fragments: Alkylidyne Because alkyl groups can form either one or two bonds to metal centers, it is not surprising that alkyl groups can also form bonds to three metal centers. Such bonding leads to the alkylidyne structure (Structure **3.66c**) (*345*) that is now

rather well known in the surface science community. This structure has been reported on Pt(111) (*346, 347*), Pd(111) (*348*), Rh(111) (*345*), and Ru(001) (*349*). Its characterization included the comparison of both its vibrational (*13*) and valence band photoemission (*12*) spectra with the corresponding spectra of the ethylidyne ligand in metal complexes. The method of preparing surface ethylidyne involves exposing an appropriate surface to acetylene or ethylene gas at room temperature. The history of its characterization on surfaces has been discussed previously (*11, 53, 350*). More recently ethylidyne has also been characterized on Pt(111) by using NEXAFS (*351a*) and on dispersed Pd catalysts by using transmission IR (*350*). Its relationship to hydrogenation catalysis also has been explored (*351b*). The presence of surface benzylidyne (Structure **3.1a**), a structure containing three metal–carbon σ bonds, was proposed to result from the methyl group C–H scission reactions of toluene (Structure **3.1b**) with Ni(111), Ni(100), Pt(100) (*351*), and Pt(111) (*69b*).

Ethylidyne was originally prepared in metal complexes by reacting the acetylene hexacarbonyl dicobalt complex with acid in a polar solvent (Scheme 3.7) (*352*). If only the hydrocarbon parts of the complexes in the reaction are considered, the reaction is analogous to that found on surfaces: the conversion of bound acetylene to ethylidyne. Although the metal complex reaction is more complex than the surface reaction—metal–metal bonds in the complexes must be broken and reformed—insight may be gained into the mechanism of the surface reaction by studying this solution phase reaction. In general, alkylidyne ligands are prepared in complexes from trihalogenated organics (R–CCl$_3$) in which the C–Cl bonds are broken to form the metal–carbon bonds in the complex. The R substituent group can be a wide range of common organic structures (*333*).

One issue that is not completely resolved about ethylidyne is whether the carbon–metal bonding is more accurately represented by three localized carbon–metal bonds or a more delocalized picture where the carbon atom is triply bonded to the center of the threefold site (*353*). For alkylidyne complexes some experimental results are consistent with the localized picture while other data indicate a more delocalized picture. For instance, mass spectral analyses of the alkylidyne noncarbonyl tricobalt complex

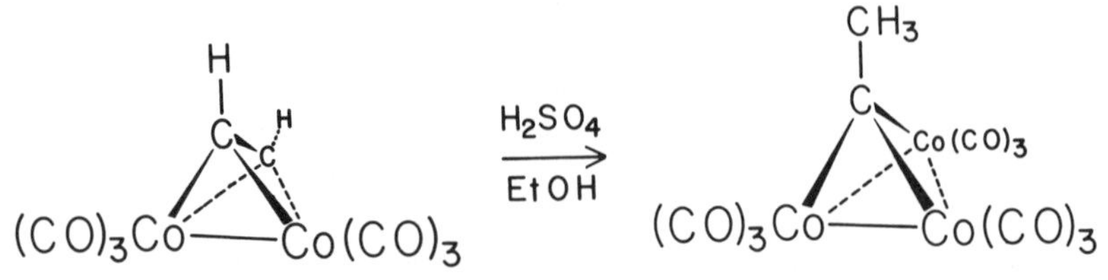

Scheme 3.7. First method used to prepare the ethylidyne nonacarbonyl tricobalt complex from the acetylene hexacarbonyl dicobalt complex.

indicate that the CO_3CR fragment is very stable and that the C–Co bonds are stronger than the Co–Co bonds (*354, 355*). On the other hand, [59]Co-NMR and [36]Cl-NMR studies (*356*) suggest an *sp*-hybridized carbon atom. More recent valence band photoemission work (*353*) also suggests that the carbon atom is *sp* hybridized. This finding indicates that the bonding is delocalized over the three cobalt atoms in the triangle. A result of this nature would be consistent with the bonding picture discussed earlier of the bridging methylene species (*183*) and leads to the general rule proposed by Chesky that the hybridization of a coordinated carbon atom is determined by the geometry of its nontransition-metal substituents only.

The large variety of alkylidyne complexes (*333*) foreshadows what may be a very interesting surface chemistry for surface alkylidyne species. Although some longer chain olefins have been studied on Pt(111) by using LEED (*357*), there have been few examples of thorough analyses of substituted olefins on surfaces and their propensity to react to form surface alkylidyne moieties. Substituents on ethylene may enhance or impede the formation of ethylidyne and may lead to insights concerning the mechanism of formation of surface ethylidyne. Long-chain olefins could possibly lead to the preparation of interesting hydrocarbon superstructures above a surface, especially if they are multifunctional and can be designed to form network bonds between surface alkylidyne moieties. Coadsorption studies are already underway for ethylidyne on dispersed palladium catalysts (*358*), and they show promise in aiding in the characterization of the interactions of coadsorbed species.

One point that might also be relevant to the formation of ethylidyne species on surfaces relates to the β-hydride elimination mechanism discussed for alkyl decomposition (*45*). The process is, as are all homogeneous catalytic processes, reversible in transition metal complexes. This statement means that a hydride ligand and an ethylene ligand can react to form an alkyl ligand in a manner that is the reverse of the reaction in Scheme 3.5. This reaction may also be similar to the surface reaction that forms ethylidyne. The only difference is the additional loss of the two hydrogen atoms on the metal-bonded carbon atom of the surface alkyl species. It is possible, therefore, that for the ethylidyne reaction on surfaces, the ultimate difference in metal complex coordination and chemisorption could be attributed to the surface proximal effect (*65*) in which the C–H bonds that have hydrogen atoms close to the surface undergo C–H bond scission.

3.15.3.4   Doubly Bonded Hydrocarbon Fragments   In contrast to the species just discussed that have metal–carbon single bonds, the carbenes are believed to have metal–carbon double bonds (*334, 359*). A metal–carbon double bond is indicated by the carbene's structure (Structure **3.66d**) in which the central carbon atom is bound to only three other atoms (a metal atom and two other ligand atoms). This situation implies that π bonding must occur so that the central carbon atom can achieve its octet of electrons. For the carbene in Structure **3.66d**, the carbon atom adjacent to the metal center has an empty p orbital just like that on the methylene bridging species (Structure **3.66b**). Intuitively, the most likely π-bonding interaction should be with the metal atom because both the alkyl species (labeled R′) and the heteroatom (labeled X), which is usually an oxygen atom, have enough bonds to achieve an octet of electrons. However, there is apparently only a small amount of π character between the metal and carbon atoms because NMR has shown that there is only a small barrier to rotation about the metal–carbon bond. Because the d-orbital energies on the metal atom are higher than those of the lone pairs on the heteroatom, the empty carbon p orbital tends to accept electron density from, and thus favor a double bond with, the heteroatom X over the metal atom (*334*). The result is partial double-bond character for the central carbon atom with both the metal atom and the heteroatom.

The majority of the methods used to prepare this type of carbene in transition metal complexes involve cations or anions in the solution phase (*334*). However, the carbene shown in Structure **3.66d** can be prepared by reacting alcohols or amines with coordinated methyl isocyanide (Scheme 3.8) in which X can be either oxygen or nitrogen; for the nitrogen case, one more R substituent would be present. This preparation could be easily mimicked on a surface because both methyl isocyanide and alcohols or amines are quite volatile. A similar reaction is also possible with alkynyl ligands (terminally bound acetylenes), but they are as yet uncommon surface structures (*334*).

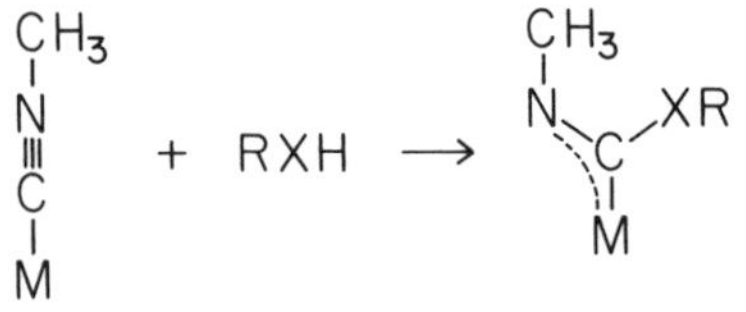

Scheme 3.8. Reaction of coordinated isocyanide with a lone-pair donor to form a coordinated carbene.

In the second type of carbene, the vinylidene class of ligand (Structure **3.66e**) (*359*) the double-bond character to the metal center is significant because the metal-bound carbon atom is not bound to atoms with available lone pairs for π bonding. The vinylidene species can be viewed as a terminally bound ligand whose bonding consists of ligand-to-metal σ donation through the sp hybrid orbital on the carbon atom, and metal-to-ligand π back donation into the unrehybridized π orbital on the carbon atom adjacent to the metal. The unusually short metal–carbon distance (1.85 Å) and the very low field NMR chemical shift of the metal-bound carbon atom (300–450 ppm) indicate that the metal–ligand bond has a bond order between two and three, which indicates that vinylidene is one of the strongest back bonding ligands (*281*).

Vinylidene ligands are prepared in metal complexes by numerous methods that have little relation to surface reactions (*359*). One of them, however, involves a 1,2 hydrogen shift reaction of a side-bonded acetylene molecule (Scheme 3.9). While it is unlikely that this method could yield a vinylidene surface species, it is conceivable that this mechanism could be involved in the conversion of chemisorbed acetylene to the surface ethylidyne species (discussed earlier). This mechanism was originally proposed by Kesmodel (*360*).

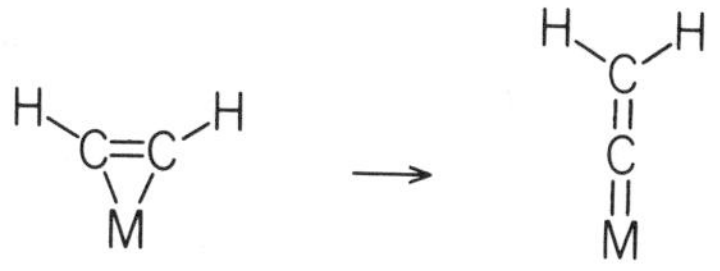

$$
\begin{array}{ccc}
\mathrm{H}\!\diagdown_{\!\!\!}\mathrm{C}\!=\!\mathrm{C}\diagup^{\mathrm{H}} & & \mathrm{H}\diagdown_{\mathrm{C}}\diagup^{\mathrm{H}} \\
\diagdown\!/ & \longrightarrow & \|\| \\
\mathrm{M} & & \mathrm{C} \\
& & \|\| \\
& & \mathrm{M}
\end{array}
$$

Scheme 3.9. Illustration of a 1,2 hydrogen shift reaction that generates a vinylidene ligand from π-bonded acetylene.

3.15.3.5   *Triply Bonded Hydrocarbon Fragments*   In addition to single and double metal–carbon bonds, the carbynes have metal–carbon triple bonds (Structure **3.66f**) (*334*). Again, there is little question of the bond order for this species because no other alternative π-bonding modes are possible. Carbynes can be prepared from carbenes by treating them with BCl_3 gas. The BCl_3 strips away the heteroatom part of the carbene and leaves behind the carbyne. Because BCl_3 is a gas, this reaction may be applicable to a surface reaction.

3.15.4   *σ-π Isomerizations of Hydrocarbon Fragments*

So far, σ-bonded hydrocarbons and π-bonded hydrocarbons such as acetylene, ethylene, and the aromatics have been discussed. Now, it seems reasonable to wonder if conversions of π-bonded ligands to σ-bonded ligands can generally occur in coordination complexes much like the β hydride elimination reaction. This situation has in fact been shown to be the case (*116*). In some reactions, a π-bound species can isomerize to a σ-bound species, and in others the converse can occur. Ligands have also been observed to rapidly oscillate between π and σ structures. These isomerizations can result from either a change on the metal center or a change on the ligand or when two structures have very similar energies (*116*).

One of the driving forces for this process, much as for other ambidentate ligands, centers around the HSAB theory. The σ-bonded organics are considered hard ligands, whereas the π-bonded organics are considered soft ligands. If the metal center that is bound to a π-bonded organic gains electron-withdrawing or π-acceptor ligands, the metal center will be hardened and σ bonding will be favored, possibly leading to isomerization. Conversely, donor ligands on the metal center could lead to the isomerization of a σ-bound ligand to a π-bound ligand. Other effects include the onset of coordination unsaturation, which occurs when a ligand dissociates from the complex. This condition could allow a σ-bonded alkyl complex to π bond to the metal atom in order to reestablish coordination saturation. Changes on a ligand that make it an aromatic species could also lead to a σ-to-π bonding transformation (116).

The σ-to-π bonding conversions could lead to some interesting surface chemistry. One good possibility may be found in the chemisorption of cyclopentadiene on surfaces. Cyclopentadiene bonds to metal centers in the side-on fashion, as discussed in the section on aromatics, and in a σ-bonded fashion (360, 362). It is possible that, at low coverages, the cyclopentadiene ring may be flat lying and, as the coverage is increased, may tend to assume the σ-bonded structure to alleviate the effects of steric crowding on the surface. A similar effect has already been reported for pyridine on Ag(111) (305). Other interesting possibilities include characterizing σ-to-π conversions on surfaces with various coverages of coadsorbates that change the electronic structure of the surface. The coadsorbate may change the hard or soft surface atom to its opposite and lead to a σ-to-π or π-to-σ isomerization.

3.16   Atomic Species

The ligands that have been discussed contain more than one type of atom to form a molecular species. There is, however, a class of ligand that contains only one type of atom; the majority are simply monatomic ligands but some are multiatomic species of only one type of atom. These ligands are grouped together in the class labeled atomic species. For the multiatomic species, this grouping is fuzzy because they bond much like diatomic and aromatic species. They are placed in this group, however, because they must be prepared from the bulk elemental material, a method that, when used for surface adsorption, could provide any one of a number of surface species including the monatomic species. This class includes the monatomics of oxygen, nitrogen, and the halogens and mono- and polyatomic species of sulfur, arsenic, and phosphorus.

3.16.1   Oxo, Nitrido, and Halogeno Ligands

One of the most common monatomic ligands is the dianion of oxygen (O_2^-) (57), the oxo ligand, which can coordinate in the terminal mode (Structure 3.70a) or the bridging mode (Structure 3.70b) in which the

M–O–M bond angle can equal 180° or less. Oxo ligands are excellent π electron donors because all the p orbitals on the ligand are filled and donate substantial electron density to the metal (Figure 3.19). This metal–oxygen π character is reflected in the unusually short M–O bond lengths found in oxo species. Because of this donor ability, oxo ligands are coordinated to metal centers that have high oxidation states, that is, species that are electron poor (*57*). Oxo ligands can be characterized by using vibrational spectroscopy; a monometal mono oxo complex has a metal–oxygen stretching frequency between 900 and 1050 cm^{-1}, and a bent bridging oxo ligand has asymmetric and symmetric stretching frequencies near 750 and 500 cm^{-1} (*57*).

$$\begin{array}{ccc} O & & O \\ \| & & \diagup \ominus \diagdown \\ M & & M \quad M \end{array}$$

$$a \qquad\qquad b$$

Structures **3.70a** and **3.70b**. Bonding modes of the oxo (O^{2-}) ligand.

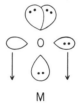

Figure 3.19. Illustration of π donation from the oxo ligand to metal center.

The nitrido ligand (N^{3-}) is similar to the oxo species but is a better π donor and is in fact the best known π donor ligand (*58*). It is also frequently found to coordinate to high-oxidation-state metals, usually the second- and third-row transition metals, because these can achieve the highest oxidation states. It generally bonds in a terminal mode (Structure **3.71a**), although various linear and planar bridging structures (Structures **3.71b** and **3.71d**) that are not directly analogous to surface structures are also known. Bent bridges (Structure **3.71c**) have been proposed in some polymeric structures, but these are considered to be unlikely (*58*). For the terminal nitrido ligand, vibrational spectroscopy has detected an M≡N

$$\begin{array}{cccc} N & & N & \\ \| \| & M-N-M & \diagup \diagdown & M-N \diagup M \\ M & & M \diagdown \diagup M & \diagdown M \\ & & N & \end{array}$$

$$a \qquad b \qquad c \qquad d$$

Structures **3.71a**-**3.71d**. Bonding modes of the nitrido (N^{3-}) ligand.

stretching frequency between 1020 and 1150 cm^{-1} and the bending mode between 300 and 380 cm^{-1} (58).

Halogeno ligands (Cl$^-$, Br$^-$, I$^-$) have a poorer π-electron donating capability than the oxo or nitrido ligands, but they have a greater ability to coordinate to several metal centers as bridging species. The terminal bonding mode (Structure 3.72a) is quite common (77), but twofold (Structure 3.72b) and threefold (Structure 3.72c) bridging species are also known (363).

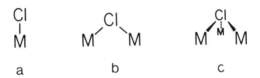

Structures 3.72a-3.72c. Bonding modes of the halogeno ligands (shown for chlorine).

The halogeno ligands are, however, difficult to characterize by using vibrational spectroscopy because metal–halide stretch peaks fall in the low-frequency region between 750 and 100 cm^{-1}; the M–F bonds lie at the higher frequency end of the region, and M–I bonds lie at the low-frequency end (77). Metal–halogen stretching frequencies increase with the oxidation state of the metal and decrease with increasing coordination (364). Both of these effects can be attributed to an M–X π-bonding interaction much like that seen for oxo and nitrido ligands because an enhanced positive charge on the metal could draw additional π electron density from the halogen and increase the M–X bond order and thus its stretching frequency. Whereas increasing the metal oxidation state is an increase in positive charge, increasing coordination increases the negative charge on a metal center and thus attenuates M–X π bonding. This situation results in a lower M–X stretching frequency. A similar effect was also seen in the dimethyl sulfoxide ligand.

3.16.2   Atomic Species Bound in Threefold Sites

Besides the halogeno, nitrido, and oxo atomic species, which infrequently bridge bond, there is a group of monatomic ligands that bond solely in a triply bridging manner (365). These species bond to the threefold bridge site of a tricobalt nonacarbonyl fragment in the manner shown in Structure 3.73. The E can be sulfur (366), arsenic (367), phosphorus (368), aluminum (369), boron (370), silicon (371), and germanium (372). This atom may contain other substituents bound to itself as in the case of boron, which has a triethylamine ligand (370), and silicon and germanium which are both coordinated to tetracarbonyl cobalt fragments (371, 372). If an atomic species cannot be placed in the threefold site, it is most likely due to its atomic size because, so far, only atomic species having van der Waals radii under 1.30 Å have been found in the threefold site. Atoms with larger atomic radii have resulted in open molecules in which three individual $Co(CO)_4$ fragments bond to the heteroatom, and the cluster is not formed (*365*).

Structure **3.73**. The tetrahedral heteronuclear cobalt carbonyl cluster. E = B, S, P, Si, Se, As, Ge, or a variety of transition metals.

One interesting aspect of these heteroatom clusters is that if the E heteroatom contains no CO ligands, then no bridge-bonded CO ligands are found in the cluster between the transition metal atoms. When E is a transition metal with coordinated CO ligands, then bridging CO ligands are present (*365*). An explanation for this phenomenon involves ligand-crowding effects (sterics) where apical CO ligands perturb the other CO ligands into an icosahedral arrangement. This argument is rather weak, however, since ligands other than CO on that heteroatom also allow bridging to occur (*365*). The effect is thus poorly understood, and at present electronic effects cannot be ruled out. It is possible that nonmetal species such as sulfur occupying the E site can, through an electronic interaction, disfavor the presence of bridging CO ligands. If this situation is the case, it could be related to the well-known poisoning effect of sulfur to the methanation reaction on surfaces (*373*). On surfaces, coadsorbed sulfur may eliminate bridging CO species in the same way that it occurs in the tricobalt cluster. If the methanation reaction requires bridging CO moieties in order to proceed, then atomic sulfur could stop the methanation reaction with the same effect that eliminates bridging CO ligands in these heteroatom clusters. A recent IRAS study on Ni(111) (*374*) reported such an effect on bridging CO by coadsorbed sulfur.

3.16.3   *Chemisorption of Atomic Species*

Many atomic species have been studied on surfaces, but, unlike the surface-bound counterparts of other ligand classes, little similarity in bonding can be noted. Surface oxygen has been shown by ARUPS to generally occupy hollow sites on surfaces (*59*); a similar result has been found for nitrogen adatoms on both Mo(111) (*60*) and Fe(100) (*375*). These results differ from the complex species, which tend to form terminally bound species. Chlorine can chemisorb at a threefold hollow as on Rh(111) (*376*), but it can also form gold chloride molecules as on Au(111) (*377*). On

Pd(111) and Pt(111) (*378*), chlorine adsorbs out of registry with the surface sites. These conclusions may be incorrect because the lack of registry could be due to antiphase domains (*379*). Fluorine has been found to form a (1 × 1) LEED pattern on Pt(100) (5 × 20) (*380*); PtF$_4$ forms at exposures exceeding a monolayer. Bromine forms either a $p(2 × 1)$ or $c(4 × 2)$ LEED pattern on Ag(110), depending on surface coverage (*381*); AgBr forms with coverages in excess of a monolayer. Sulfur is really the only species in this group that bonds on surfaces as it does in complexes, in the hollow sites (*382, 383*).

3.16.4   Polyatomic Elemental Species of Sulfur, Phosphorus, and Arsenic

This class of atomic species includes not only the monatomic ligands, but also the polyatomic ligands, which include disulfur (S$_2$$^-$), and dimers, trimers, and tetramers of phosphorus and arsenic. These species have not yet been reported on surfaces, but because they are usually prepared with the vapor of the elemental species S$_8$, P$_4$ or As$_4$, it is likely that similar structures could be prepared on surfaces by using elemental vapors. Disulfur can coordinate to metal centers in a variety of modes because it has as many as six lone electron pairs that can coordinate in a variety of combinations (*384*). The disulfur coordination modes that may have analogues on surfaces include the one-metal (Structure **3.74a**) and two-metal (Structure **3.74b**) side-on bridging structures, the monosulfur bridge (Structure **3.74c**), and the dimetal side-on bridging mode (Structure **3.74d**) common in acetylene coordination chemistry. The disulfur ligand linearly bound to one metal center, the analogue of CO bonding, is unknown in transition metal complexes (*384*). The disulfur stretching frequency falls between 600 and 450 cm^{-1}, but the frequency is not characteristic of the bonding mode. Disulfur ligands can be prepared from S$_8$, S$_2$Cl$_2$, solution-phase dianions, or monosulfur compounds such as hydrogen or sodium sulfides (H$_2$S, Na$_2$S) (*384*).

The multiatom ligands of phosphorus and arsenic are based on the tetrahedral structure (*81*) of elemental phosphorus (P$_4$) (*385–387*) and arsenic (As$_4$) (*388*). For both elements, it is possible to substitute a tricarbonyl cobalt fragment [Co(CO)$_3$] for one or all of the atoms in the

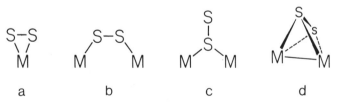

Structures **3.74a-3.74d**. Bonding modes of the disulfur ligand.

tetrahedron. This substitution can result in the monatomic phosphorus or arsenic structure shown in Structure **3.73** or the di- or triatomic species shown in Structures **3.75a** and **3.75b**, respectively. The tetrahedron itself can also bond intact to a metal center (Structure **3.75c**) (*387*) and form a phosphorus-metal σ bond. Coordination of the tetrahedron shortens the bonds for arsenic dimers and trimers and lengthens them for the phosphorus tetrameric species, resulting in trigonal pyramidal structures (*387*).

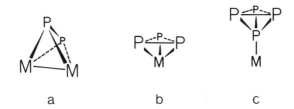

Structures **3.75a-3.75c**. Bonding of elemental phosphorus with transition metals. The phosphorus or the metal can occupy the apices of the tetrahedron in any combination. A similar situation exists for arsenic.

The structures resulting from interacting S_8, P_4, or As_4 vapor with surfaces have not as yet been investigated. The most likely result of chemisorbing P_4 on surfaces, if complete dissociation to atoms does not occur, could be the tetrahedral ligand where no cluster scission is involved; experimentation would be required to resolve this issue. S_8 may also interact nondissociatively. White phosphorus (P_4) boils at 280 °C, S_8 boils at 444 °C, and arsenic boils at 613 °C (*388*); thus it would not be difficult to deliver these species to surfaces so that their surface structures could be characterized.

3.17 Conclusion

This chapter has illustrated the diversity of ligands commonly used by the coordination chemist by arranging them into a series of 13 classes that are delineated by the commonality of their coordination bonding modes. The characteristics of each ligand have been discussed so as to provide a basis for surface scientists to gain an initial exposure to the field of descriptive organometallic and coordination chemistry. Exemplary surface chemisorption studies were discussed as a means of illustrating the qualitative similarities between coordination chemistry and chemisorption.

Although few direct similarities have been conclusively documented between organometallic chemistry and surface chemisorption, it is clear that issues now being pursued within the bounds of surface science research are very much like the issues being investigated in the field of organometallic chemistry. Issues such as the nature of ligand bonding to metal sites, characterization of new bonding modes for ligands on metal sites and delineation of the reactions of small molecules in the presence of metal atoms are not new issues pioneered by the surface scientist but have been the subject of intense investigations by organometallic chemists. It seems apparent that most any program of surface science research could benefit greatly by using organometallic chemistry as a reference point from which to gain understanding into the more likely results that may be expected in a given surface system and to gain insight into the results that are unique to surfaces.

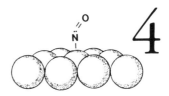

Bonding Sites in Clusters and Their Relationship to Surface Bonding Sites

ALTHOUGH THE WIDE VARIETY OF LIGANDS found in the organometallic literature offers a broad range of potential systems to be studied on surfaces, the question remains whether or not a surface can be accurately modeled by a small cluster of metal atoms or, indeed, by a single metal center that is coordinatively saturated by ligands. This model has been criticized because, although the electronic structure of single metal atoms surrounded by ligands can be described theoretically by a series of discrete orbitals, many current views of the electronic character of the surface involve the more complicated band structure theory (23). The added complexity of the band structure leads to a more complex picture of the metal–ligand interaction than that found in metal complexes.

An unfortunate consequence of this attitude is that the descriptions of ligand–metal bonding in complexes and clusters that are based upon aspects of orbital symmetry and overlap are not generally extended to theoretical studies of surface systems. These bonding descriptions could be used for surface bonding if it could be established that the band structure of a surface could be divided into its components of given symmetries with respect to a surface site. Then it would not be difficult to describe the surface–adsorbate bond in terms of the overlap of adsorbate orbitals and surface bands that have the same symmetry in direct analogy to the metal–ligand bonding interactions in metal complexes.

Within the field of organometallic chemistry a class of theoretical calculation generates this type of metal-ligand bonding description, which seems, even in the absence of band theory considerations, able to enhance the understanding of adsorbate–surface bonding. This chapter will discuss these calculations for the relevant bonding sites in clusters that also appear on surfaces and show how they can be applied to gain a better understanding of surface chemisorption. The next chapter will discuss how band theory considerations can extend this type of calculation to surface chemisorption interactions (24, 25).

4.1   Frontier Molecular Orbital Theory

The class of theoretical calculation to be discussed in this chapter uses the frontier molecular orbital (FMO) approach to the study of bonding (389). FMO calculations generally provide information concerning the relative energy levels and corresponding point-group symmetries of the orbitals of molecular fragments, that is, parts of molecules, as opposed to complete molecules. By making a judicious choice of two fragments and then placing the two fragments in a bonding geometry, it is possible to study the nature of the orbital overlap of the two fragments and to determine if bonding can occur and, if so, the specific orbital overlaps that lead to the bonding.

For metal complexes and clusters, the FMO approach is a very convenient method for studying the bonding interactions of metal centers and ligands, because a metal complex can naturally be divided into a ligand fragment and a metal complex or cluster fragment that is missing one ligand. The bonding of the ligand to the metal can then be understood in terms of orbital overlap by way of FMO calculations on the relevant fragment–fragment molecular system that will be composed of a ligand fragment and a metal complex or cluster fragment that contains a particular coordination site.

The resulting interactions predicted by these FMO calculations will be shown to bear a significant likeness to the kinds of interactions that occur at vacant surface sites, in spite of the fact that the calculations do not involve the band structure of the surface. The orbital structure of the metal complex or cluster fragment at its vacant bonding site will be shown to be a model of the orbital structure of the corresponding surface site. The bonding interactions of a ligand in that metal complex or cluster site, therefore, could be looked upon as at least a first approximation of the bonding interaction of that ligand with the corresponding metal surface site. The metal sites in complexes and clusters studied in this manner include the monometal (390), twofold bridge (391a), threefold triangular (43) and long-bridge (391b) sites. Because these sites are directly analogous to the sites commonly found on surfaces, it is not unlikely that their orbital structures will also be like those on surfaces. The ultimate correctness of this analogy will depend upon the inclusion of band theory into these models; but, as will be shown in the next chapter, early successes of this type have already been achieved (*24, 25*).

In this chapter, the FMO approach to ligand bonding in metal complex and cluster bonding sites will be reviewed. These calculations will be presented as a means of illustrating the kinds of qualitative insights into metal–ligand bonding that can be gained from calculations that include the symmetries of metal orbitals and also how they can be extrapolated to provide insight into various aspects of surface–ligand bonding. Several calculations of this nature that are centered on a particular ligand have already been mentioned in earlier sections of this work. The calculations for the bonding of diatomic ligands (*104*) and the CO_2 (*203*) ligand fall into this category. The calculations chosen here for illustration, except for the contributions of Lauher (*39, 393*), come exclusively from R. Hoffman's group at Cornell University (*389*), including the contributions of M. Elian (*390*), D. L. Thorn (*391a*), B. E. R. Schilling (*43*), and D. M. Hoffman (*391b*). A number of alternate calculations are also found in the literature (*392a, 392b*).

4.2   Atop Site on Naked Metal Clusters

One of the early attempts to correlate the orbital structure of a cluster to that of a surface is found in the work of Lauher (*39, 393*). By using extended Hückel calculations (EH), Lauher was able to develop a delocalized molecular orbital bonding picture for small (<15 atoms) naked metal clusters that approached a band structure as the number of metal centers grew large. Unlike band theory calculations, however, the results were derived predominantly from orbital symmetry and nodal patterns and used relative energy values only. The advantage of this calculation is that it can be applied to clusters containing atop sites whose environments mimic the atop sites on (100) and (111) surfaces and thus provide a possible orbital model of those sites on surfaces.

Lauher's method may be described by using a three-atom triangular cluster such as $Fe_3(CO)_{12}$ as an example (*39*). For the simple triangular cluster of iron atoms, 27 valence atomic orbitals are mixed to form the cluster orbitals, that is, five *d*, three *p*, and a single *s* orbital on each metal atom. When these 27 levels are combined, 27 molecular orbitals are generated (Figure 4.1). Twenty-four of the levels have energies near those of the initial atomic orbitals, but three have energies substantially higher and are predominantly metal–metal antibonding. Empirically, it has been shown that a level is antibonding if its energy lies above that of an isolated metal *p* orbital. From this calculation, it was concluded that when the 24 lower energy levels, called cluster valence molecular orbitals (CVMOs), are all filled, a stable cluster is obtained. For the case of the $Fe_3(CO)_{12}$ cluster,

24 of the 48 valence electrons come from the three iron atoms, 8 electrons from each, and the remaining 24 electrons are provided by the two-electron donation of the 12 CO ligands. Simple electron counting shows that this result is totally consistent with the 18-electron rule discussed in Chapter 2. The advantage of this formalism over the 18-electron rule lies in the fact that as the clusters become larger, the 18-electron rule breaks down, whereas Lauher's molecular orbital approach accurately predicts the cluster bonding.

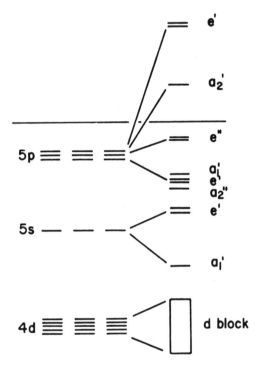

Figure 4.1. Molecular orbital picture for the transition metal orbitals for the M_3 triangular cluster. The top three are considered antibonding. (Reproduced from reference 39. Copyright 1978 American Chemical Society.)

By using Lauher's delocalized approach (39) to cluster bonding, the bonding of the CO ligands in a delocalized fashion can also be described. As seen from Figure 4.1, the s- and p-derived molecular orbitals lie above the d block of molecular orbitals and compose the acceptor levels for the CO ligand bonding. The d block of orbitals accommodates the metal-derived electrons. Each of these acceptor levels has a well-defined point-group symmetry with respect to the cluster. It is inappropriate to consider each CO bond as an interaction of these delocalized acceptor levels with a localized 5σ orbital on one CO ligand. Instead, a linear combination of the

12 5σ CO donor orbitals is made. This combination results in the formation of 12 filled orbitals that are centered on the CO ligands and that have symmetries identical to those of the acceptor levels on the three-atom cluster. The ligand–metal overlap then is composed of the 12 delocalized acceptor levels on the metal cluster; they accept electron density from the 12 delocalized donor orbitals on the CO ligands (*39, 394*).

An interesting aspect of this result is that it contains the rudiments of the FMO band theory calculations for surfaces. Even though there are few enough levels in the cluster to be resolved in a discrete fashion, the 5s- and 5p-derived acceptor molecular orbitals become dispersed over an energy range. This situation is analogous to what occurs on metal surfaces when a continuum of levels forms a band. That the CO bonding in these delocalized levels in the cluster can still be understood suggests how CO bonding to the surface band structure can be better understood in terms of the orbital overlap and symmetry considerations found in discrete clusters (to be discussed in detail later in this section).

Lauher's delocalized bonding description can also be successfully applied to larger clusters such as octahedra, dodecahedra, icosahedra, etc. (*39*) and produce results similar to those found for the triangular cluster. In each case, the molecular orbitals of the metal fragment that lie below the energy of the isolated metal p orbital are bonding, and those that lie above are antibonding (high-lying antibonding orbitals, HLAO). A compendium exists for a wide variety of these clusters (*39*) and delineates the number of CVMOs, HLAOs, and valence electrons for stable cluster bonding.

By using these larger clusters, it is possible to better understand the bonding on specific surface sites, because metal atoms on certain of these larger clusters occupy environments similar to the environment of an atom on a given crystal habit plane (*393*). As shown in Figure 4.2, a metal atom lying in a (100) habit plane resides in an environment similar to the metal atom labeled G in a square pyramidal cluster (Figure 4.2a). Although the base of the pyramid is a truncated version of a (100) surface, the G atom has the appropriate set of nearest neighbors for a metal atom in a (100) surface, an environment that is adequate to gain at least qualitative understanding of a metal atom on a (100) surface. In a similar manner, the atom labeled K in the 20-atom tetrahedron (Figure 4.2b) occupies a position analogous to that of a metal atom on a (111) surface (*393*).

From calculations on these bare metal clusters, it is possible to determine the number of CVMOs associated with a given metal site from the calculated atomic orbital coefficients of each CVMO. For a cluster such as the M_3 species discussed earlier, all the atoms are equivalent and thus each has the same number of CVMOs. For larger clusters, however, the number can vary from site to site but can be determined for each site. Metal sites with a high coordination of metal atoms have fewer CVMOs, whereas metal sites with a lower coordination have more CVMOs. For the (100) model

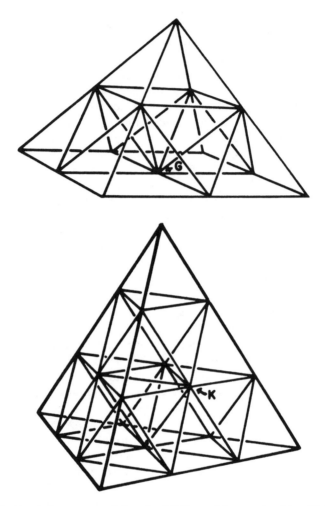

Figure 4.2. Metal clusters containing the (100) model atop site (site G) (top) and the (111) model atop site (site K) (bottom). (Reproduced from reference 393. Copyright 1979 American Chemical Society.)

site (Figure 4.2a), there are 6.21 CVMOs, whereas for the (111) model site (Figure 4.2b), there are 5.98 CVMOs (393).

Given an average value of six orbitals per model surface site, for a cluster of nickel atoms, the metal atom in the model surface site would place its 10 valence electrons in five of these orbitals. One orbital would, therefore, remain empty and thus would be able to accept a lone pair of electrons from a CO molecule or other ligand (393). The calculations show that this acceptor orbital for both the (100) and (111) sites is composed predominantly of the p_z orbital and some contribution from the s and d_{z^2}

orbitals (*395*). This finding indicates that the acceptor level extends normally along the *z*-axis on the model surface and so facilitates overlap with the donor orbital of CO.

Because this acceptor orbital is predicted to exist on model (100) and (111) sites, most likely an orbital of this nature or something similar to it should be present on the atop surface sites on (100) and (111) surfaces. This possibility is likely because there must be an orbital of appropriate symmetry to overlap with the 5σ level of chemisorbed CO. This lone orbital lying normal to the model surface in both of these cluster systems is of a_1 point-group symmetry, which is the case for the 5σ level of CO bound normal to the surface. This level should be kept in mind for the discussions to follow because it will provide a link between the metal complex calculations to be discussed next and analogous surface interactions.

4.3 Bonding Sites in Organometallic Compounds

The existence of Lauher's work on naked metal clusters is an important theoretical bridge between surface and cluster chemistry. Although the FMO calculations are carried out solely for organometallic systems, when considered carefully, they can offer new qualitative insights into bonding and reactivity of ligands on surfaces. These new insights are possible because the metal complex fragments studied in these calculations have ligand bonding sites that correspond almost exactly to the variety of bonding sites, for example, the atop, bridge, and threefold sites, that are commonly found on single-crystal surfaces. By understanding the interactions of ligands with these metal complex and cluster sites, it becomes possible to consider their applicability to surface bonding and reactivity.

In general, these sites can be generated by considering a complex or cluster that contains the appropriate site and that has a known molecular orbital structure (*43, 390, 391*). By theoretically removing appropriate ligands in the complex, the site under consideration will be exposed. Perturbation theory can then be used to determine how the molecular orbital structure of the remaining fragment has been changed with regard to both orbital relative energies and symmetry. Most importantly, energies and shapes of the orbitals that lie in the vacant site of the complex can then be determined. These are the levels that accept or donate electron density from or to the orbitals of the ligand species that occupy the site and offer insights into ligand–metal interactions. These sites have ligands as nearest neighbors and not metal atoms, as is the case for surfaces. However, the central issue to be discussed in this chapter is one of symmetry, which should be qualitatively maintained regardless of the nearest neighbors.

These same metal complex and cluster sites for ligand bonding can also be generated by joining appropriate smaller fragments and distorting them, if necessary, to arrive at the appropriate fragment. During this process, the energies and symmetries of all the levels can be monitored until the desired fragment is obtained. Without exception, both this method and those described earlier are found to yield the same orbital picture of a given bonding site.

4.3.1   The Monometal or Atop Site

The simplest metal site that can be described by an organometallic fragment is the terminal (on-top) site (390). This monometal site can be generated easily from the common ML$_6$ complex of octahedral symmetry (Figure 4.3). As discussed earlier, the metal complex of octahedral (O$_h$) symmetry has split $5d$ orbitals, which are shown on the left of Figure 4.3. If one ligand is removed from the z-axis of the complex, then the symmetry of the complex is reduced to C$_{4v}$, and the d_{z^2} orbital is significantly stabilized (Figure 4.3, right). This situation results in a low-energy acceptor level of a_1 symmetry that is appropriate for overlap with two electron-donor species. The three lower energy levels, which provide the electron density for π back bonding interactions, are shifted only slightly in energy although their symmetries are reduced from t_{2g} to e for the d_{xz} and d_{yz} levels, and to b_2 for the d_{xy} level. The resulting acceptor lobe is highly directional along the z-axis because it incorporates not only the d_{z^2} level, but the

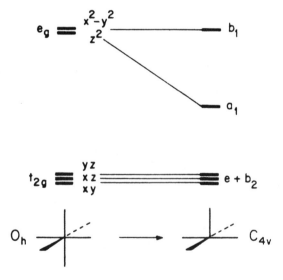

Figure 4.3. Variation of the relative energies of the d orbitals in an octahedral crystal field when one of the ligands is removed to expose a monometal atop site. The a_1 level is the new acceptor orbital. (Reproduced from reference 390. Copyright 1975 American Chemical Society.)

valence s and p_z levels as well. In octrahedral symmetry the d_{z^2}, p_z, and s levels have symmetries of e_g, t_{1u}, and a_{1g}, respectively. Because each level has a different symmetry, they cannot mix with each other. However, in C_{4v} symmetry, the symmetry of the monometal site, they all have a_1 symmetry and thus can mix together as shown in Scheme 4.1. The result is the formation of a highly directional acceptor lobe within the vacant site. Similar reasoning can be applied to obtain a similar result for the vacant site in a monometal complex fragment of C_{3v} symmetry (Figure 4.4) (*390*).

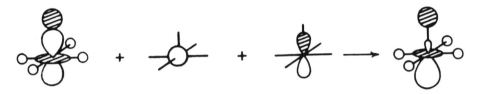

Scheme 4.1. Orbital diagram of the combination of the d_{z^2}, s, and p_z metal orbitals of a metal center in a complex of C_{4v} symmetry. This combination produces the low energy a_1 acceptor lobe for the atop site. (Reproduced from reference 390. Copyright 1975 American Chemical Society.)

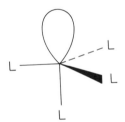

Figure 4.4. Schematic of the low-energy acceptor lobe in a complex fragment of C_{3v} symmetry.

These two results are analogous to the results of Lauher's work for the model of a surface site as derived from a naked cluster of metal atoms. The atop site of a (100) surface is of C_{4v} symmetry whereas the atop site of a (111) surface is of C_{3v} symmetry. Both have a single vacant orbital whose lobe extends normally to the vacant site composed of a mixture of d_{z^2}, p_z, and s metal orbitals. Most likely, therefore, the qualitative results obtained from studying the interaction of this monometal site in a complex with other ligands can provide qualitative insight into analogous interactions on metal cluster sites of similar symmetry and possibly on surfaces. The issue of surface band structure will be added to this argument in the following chapter.

The monometal atop site has been used in the study of dihydrogen bond scission in monometal complexes (*24*). H_2 scission plays a significant role in a variety of homogeneously catalyzed reactions, and understanding the optimum approach geometry for the dihydrogen–metal site interaction is fundamental to catalytic phenomena on surfaces that involve hydrogen. In the following chapter, it will be shown that the interaction to be discussed next is similar to an analogous series of interactions occurring at a surface site.

For dihydrogen scission to occur, the bond order of the dihydrogen molecule must be reduced to zero. This condition can occur by either fully depopulating the filled σ-bonding orbital or by fully populating the σ^* orbital. Neither of these processes can be fully realized and thus neither can lead to H–H scission independently; both must, therefore, act in concert. This means that in any scisson process, electrons must flow from the σ bond of the hydrogen molecule to the metal atom, and d electrons must flow from the metal atom to the σ^* level of the dihydrogen ligand. To achieve this result, the dihydrogen ligand must sit in the atop site (because many homogeneous catalysts are monometal) in an optimized geometry that should be determinable theoretically (24).

The two most likely geometries are shown in Figure 4.5. The first is labeled the perpendicular geometry in which the dihydrogen is colinear with the metal center (Figure 4.5, left), and the second is labeled the parallel geometry in which the molecule is essentially side-on bound to the metal center (Figure 4.5, right). By studying the possible orbital overlaps of these two geometries, the most stable geometry can be determined.

The molecular orbital pictures for the two geometries are shown in Figure 4.6. The perpendicular geometry is shown on the left and the parallel geometry is shown on the right. The center of the figure shows the molecular orbital structure of the ML_5 C_{4v} fragment shown earlier in Figure 4.3. In the perpendicular geometry, both the σ_g and the σ_u orbitals of the

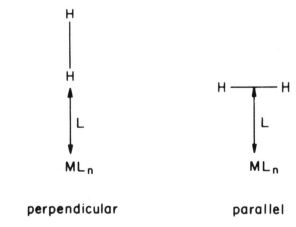

Figure 4.5. Two possible coordination geometries of dihydrogen. (Reproduced from reference 24. Copyright 1984 American Chemical Society.)

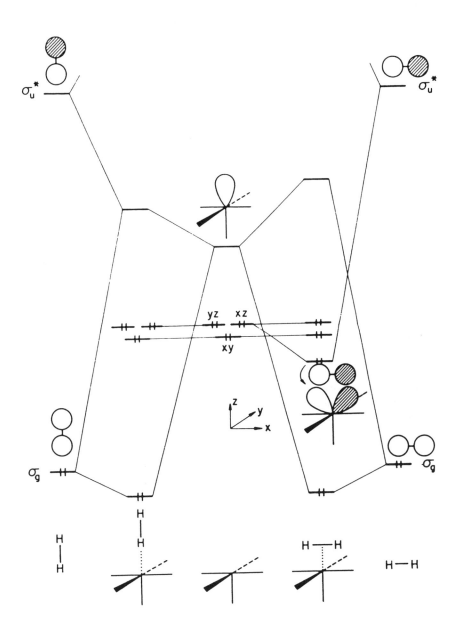

Figure 4.6. Orbital energy diagram of the bonding interactions of the perpendicular (left) and parallel (right) geometries of dihydrogen with the monometal site in a metal complex. (Reproduced from reference 24. Copyright 1984 American Chemical Society.)

dihydrogen ligand are of a_1 symmetry in the C_{4v} point group. They both mix with the a_1 acceptor lobe in the three-orbital overlap pattern and produce three new bonding orbitals (Figure 4.6, left) (one is very high in energy and is not shown). Because only the lowest orbital of the three is filled, a small bonding interaction occurs (24).

This interaction is not, however, as strong as that which results in the parallel geometry (Figure 4.6, right) (24). In this geometry, the σ_g level is still of a_1 symmetry, but now the σ_u level is of e symmetry. This geometry allows the filled σ_g to overlap with the empty a_1 donor on the ML$_5$ fragment and also allows the empty σ_u level to overlap with the filled d_{xz} level, also of e symmetry, on the ML$_5$ fragment. This latter interaction is forbidden in the perpendicular geometry case by symmetry. This situation results in two bonding interactions in which one of the bonds is a σ ligand to metal electron donation, and the other is a metal-to-σ^* π-type electron back donation. Energetically, these interactions make the parallel geometry the more stable orientation and also provide the proper impetus for H–H bond scission to occur. It seems likely, then, that H–H bond scission occurs through parallel bonding on monometal complexes. This argument is strengthened by the existence of a monometal complex that has been reported to exhibit this geometry for a dihydrogen ligand (396). It is, therefore, possible that an interaction similar to this one may also occur on a surface where dihydrogen undergoes H–H bond scission. This is especially likely because the orbital overlaps in the bonding interaction of dihydrogen are very much like that of CO. In the next chapter this situation will be shown to be the case.

Monometal complex fragments can also model an on-top site somewhat like a ridge site on an fcc (110) surface. The fragment is of C_{2v} symmetry and is shown in Figure 4.7 (390). The metal fragment and resulting vacant metal site are generated by removing two adjacent ligands from an octahedral ML$_6$ complex. The result is two empty acceptor orbitals, similar to those discussed for the ML$_5$ fragment. A linear combination of these two

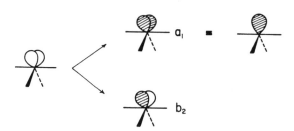

Figure 4.7. Orbital structure of a monometal site in a cluster that models a ridge site on an fcc (110) surface. At left are two individual acceptor orbitals that mix to form the a$_1$ and b$_2$ acceptor orbitals. (Reproduced from reference 390. Copyright 1975 American Chemical Society.)

acceptor orbitals results in an a_1 level of higher energy, which is the empty acceptor for orbital complexes of this type, and a filled b_2 level of lower energy, which will be the highest occupied level of the fragment (Figure 4.7). The filled b_2 level is derived from the d_{xz} level of the C_{4v} ML$_5$ fragment discussed earlier and is the π back donor orbital of the site. The significant down bending of the two ligands in the xz plane destabilizes this d_{xz} level and leads to the fragment's distinctive chemistry.

The scission of dihydrogen has been studied on this C_{2v} monometal fragment, and the results can be correlated with this destabilization of the d_{xz} level (24). For perpendicularly bound dihydrogen, the interaction is net repulsive. The orbital overlaps are similar to those for the C_{4v} metal fragment site, but now the filled d_{z^2} level adds additional repulsion to the axial overlap of the metal-σ^* bond. For parallel bound dihydrogen, however, for which the axis of the dihydrogen lies in the plane of the back bent ligands, the interaction is strongly bonding; the stability is three times that of the parallel geometry in the C_{4v} system. Qualitatively, the parallel bonding mode is the same as that discussed in the C_{4v} case. The σ bond on the hydrogen donates to the empty a_1 acceptor level on the fragment, whereas the filled d_{xz} level on the fragment back donates into the hydrogen σ^*. For the C_{2v} fragment, because the down bent ligands destabilize the d_{xz} level, the electron donation into the σ^* level is increased and thus the back bonding is greatly enhanced. This situation results in the stronger bonding interaction (24).

Much insight concerning the fcc (110) atop site comes from the behavior of the d_{xz} orbital in the C_{2v} complex fragment. Although the symmetries of the fcc (110) atop site and the complex fragment are both C_{2v}, the coordination about the metal center is four in the fragment and seven in the fcc (110) surface. The significant similarity between the two cases, however, is that some of the coordination of the metal center perpendicular to the ligand–metal bond axis is absent. This characteristic leads to the destabilized d_{xz} level in the complex fragment. A similar effect may occur on an fcc (110) surface site. A similar effect may also be present on surface sites of lower coordination (edges, adatoms, etc.) and provide a qualitative interpretation of the added reactivity of these low coordination sites; that is, back bonding interactions can be enhanced by a destabilization of the d orbitals involved in the back bonding.

4.3.2   The Twofold Bridging Site

In addition to terminal sites, it is also possible to study twofold bridging sites by using the FMO approach to ligand bonding (391). Such a study has been performed for the M$_2$(CO)$_6$ fragment (Structures 4.1a and 4.1b). M$_2$(CO)$_6$ is a rather common fragment in the organometallic literature and appears in dimeric systems such as the alkyne hexacarbonyl dicobalt complex (Structure **4.1b**). The fragment's orbital structure can be derived by either of the two methods described in the section on bonding sites in organometallic compounds. It can be derived by removing the four adjacent ligands illustrated in Scheme 4.2; the four resulting ligand-directed levels transform as $A_1 + B_1 + B_2 + A_2$, and when they are linearly combined, four molecular orbitals of 2_{a1}, b_2, a_1, and a_2 symmetries result, as shown in Figure 4.1 (*391a*).

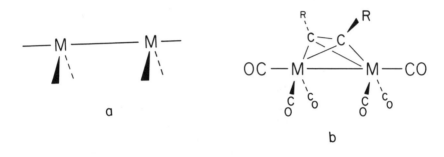

Structures **4.1a** and **4.1b**. (a) The structure of the metal complex fragment containing the bridging site. (b) The alkyne hexacarbonyl dicobalt complex with acetylene π bound in the bridge site.

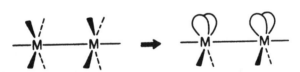

Scheme **4.2**. Illustration of the generation of the metal complex bridge site by removing four adjacent ligands and exposing four individual acceptor orbitals. Linear combination of these orbitals generates the orbital structure of the bridging site. (Reproduced from refrence 391a. Copyright 1978 American Chemical Society.)

The fragment bridge site can also be generated by first joining two $M(CO)_3$ fragments through the two metal centers and then bending the resulting fragment (which has the structure of a staggered ethane molecule) into the familiar sawhorse configuration. The resulting molecular orbital picture is the same as for the method in the previous paragraph, except for the added low-lying $1a_1$ level that is predominantly M–M σ bonding and would not be expected to appear in the calculation by the first method (*391a*).

In the resulting fragment in which the metal centers are cobalt atoms, the lower three levels (the $1a_1$, the b_1, and the a_2) (Figure 4.8) are filled,

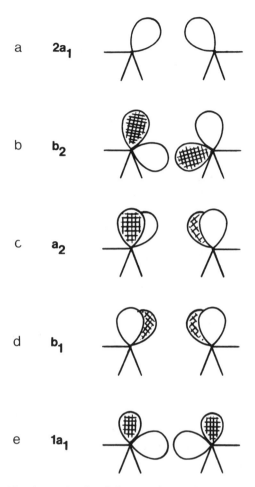

Figure 4.8. (a–d) The four orbitals of the metal complex bridge site. a and b are acceptor levels, whereas c and d are levels of appropriate symmetry for back donation. (e) The metal-metal bond orbital which is not involved in ligand bonding.

whereas the upper two levels (the b_2 and the $2a_1$) are empty. The b_1 and a_2 levels are metal–metal π bonding and antibonding, respectively. They are of appropriate symmetry to serve as donor orbitals for π back bonding. The b_2 and $2a_1$ levels are metal–metal σ antibonding and bonding, respectively. They are of appropriate symmetry for being the acceptor levels of the bridging site.

This bridging site can be related to Lauher's work on naked metal clusters in a manner similar to that done for the model atop site in monometal complexes. For the bridging case, the lobes on two adjacent naked cluster sites must be considered. It is assumed that all of the (100) sites have the same vacant acceptor lobe that lies normal to the model surface. A simple linear combination of two orbitals on adjacent sites of an extended naked cluster will generate two new levels, each with a lobe on either metal atom. One of the levels will have lobes of the same parity, whereas the other will have lobes of opposite parity (41). Comparison of this result with the b_2 and $2a_1$ acceptor levels of the bridge site shows that the two orbitals have a similar lobe structure (Figure 4.8). This finding indicates that the bridge site in the complex is probably a fair model of a surface bridging site.

A weakness in this argument is that the bridge site in the complex has no coordination in the surface plane of the two metal atoms. Therefore, the metal back donor levels may be destabilized in the complex in much the same way as was found for the C_{2v} complex fragment that modeled the (110) atop site. Although this situation is likely, the destabilization for the monometal site led only to a quantitative difference in the bonding inter-action and thus can be deemphasized in the discussion of the bridge site.

The coordination of a ligand to the bridge site can be exemplified by the bridge bonding of the acetylene ligand as described for the dicobalt system in Figure 4.9 (391a). On the left side of Figure 4.9 lie the five orbitals of the complex fragment containing the bridging site arranged according to their relative energies, whereas on the right side of Figure 4.9 lie the two filled π orbitals and the two empty π^* orbitals of acetylene, also arranged by energy. The degeneracies of the two sets of the acetylene levels have been lifted because the C–H bonds in the ligand fragment have been bent out of the C–C axis to reflect the ligand's structure when bound in the complex. When the two fragments are brought together, ligand-to-metal donor bonding results by the overlap of the filled a_1 and b_2 levels on the acetylene ligand (the π orbitals) with the empty 2_{a1} and b_2 levels on the bridging site. Metal-to-ligand back donation occurs through the overlap of the filled a_2 and b_1 levels on the bridging site with the empty a_2 and b_1 levels on the acetylene ligand (the π^* levels). The result is a very stable com-plex with a low-binding-energy molecular orbital spectrum, which is shown in the center of Figure 4.9. Several photoemission studies of these com-plexes (11, 397, 398) report low-binding-energy ultraviolet photoelectron spectroscopy (UPS) spectra of these complexes that are consistent with this orbital picture.

Most likely, a similar set of bonding interactions occurs for the bonding of acetylene in a bridging site on a surface. However, in the complex, the two filled a_1 levels repel each other to a considerable degree. The complex alleviates this problem by distorting the $Co_2(CO)_6$ fragment out of the sawhorse configuration (Figure 4.10) and making the $1a_1$ metal–metal bonding level literally a bent metal–metal bond. This distortion should be unlikely on an unreconstructed surface, and it raises the issue of the stability of this mode of bridging acetylene on a surface bridging site.

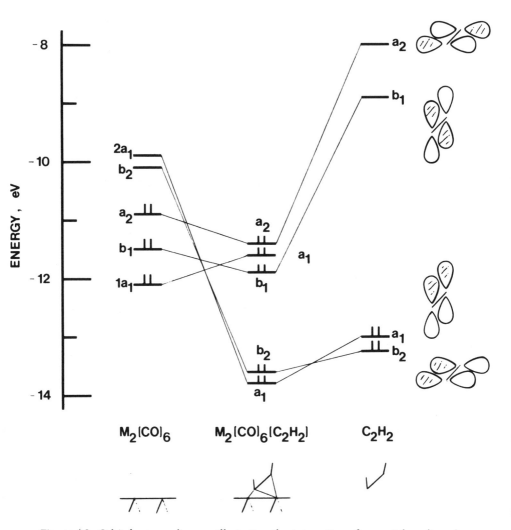

Figure 4.9. Orbital energy diagram illustrating the interaction of an acetylene ligand with the metal complex bridging site. (Reproduced from reference 391. Copyright 1978 American Chemical Society.)

When the complex is composed of two iron atoms instead of two cobalt atoms, the bridge site will have two fewer electrons. This condition will result in an additional empty level on the bridging site and lead to a somewhat different bonding picture (*391a*). The a_2 level on the bridge site is the most likely level to be vacated because it is the only filled level that is metal–metal antibonding. This situation is consistent with the smaller metal–metal spacing in the iron complex relative to the cobalt complex. This statement means that the acetylene-to-bridge site bonding will be like

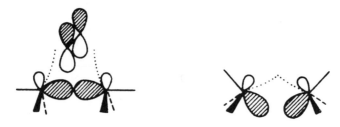

Figure 4.10. Orbital picture of the repulsion of the 2_{a1} and 1_{a1} levels leading to the bent metal-metal bond in the alkyne hexacarbonyl dicobalt complex. (Reproduced from reference 391a. Copyright 1978 American Chemical Society.)

that of the cobalt complex, except that one of the metal-to-ligand back bonding interactions will be vacant.

In addition to the diminished degree of back bonding, the empty a_2 level leads to two more subtle consequences. First, because the a_2 level is a low-lying lowest unoccupied molecular orbital (LUMO), the resulting complex is likely to undergo a second-order Jahn–Teller distortion. Calculations show that the most likely distortion is a rotation of about 25° of the C–C axis toward parallelism with the Fe–Fe bond (Figure 4.11a) (*391a, 399*). This distortion, however, does not occur. Instead, one of the Fe(CO)$_3$ fragments rotates 180° out of the sawhorse structure. This rotation destabilizes the empty a_2 level (Figure 4.11b) by diminishing the total bonding

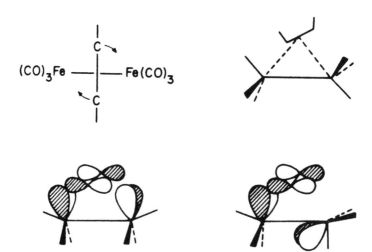

Figure 4.11. Two possible second-order Jahn–Teller distortions of the alkyne hexacarbonyl di-iron complex: (a) rotation of the C-C axis toward parallelism with the metal-metal bond (top) and (b) 180° rotation of one Fe(CO)$_3$ fragment to destabilize the a_2 level (bottom). (Reproduced from reference 391a. Copyright 1978 American Chemical Society.)

overlap of the a_2 level of the bridge site with the a_2 level of the acetylene ligand. This distortion found in the complex cannot occur on a surface; it would be interesting to learn if the former distortion, the rotation of the C–C axis, occurs on a surface site of similar electronic structure.

The second consequence of the empty a_2 level may be related to the ability of the di-iron complex to oligomerize acetylene (*391a*). As shown in Figure 4.12, a filled π level of free acetylene can easily overlap with the empty a_2 level to form, in a concerted fashion, an Fe–C bond and a C–C bond. Such bonding could lead to many structures composed of several acetylene molecules. On surfaces that are active to oligomerizing acetylene, a similar kind of orbital overlap may be involved in the mechanism of the surface reaction.

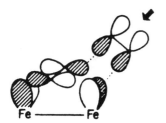

Figure 4.12. Overlap of the empty a_2 level with π orbital of an incoming acetylene molecule. This interaction is believed responsible for the catalytic activity of the di-iron complex in acetylene oligomerization. (Reproduced from reference 391a. Copyright 1978 American Chemical Society.)

4.3.3   The Threefold Site

Threefold triangular or hollow sites are also found in organometallic complexes and have also been the subject of FMO calculations (*43*). Threefold hollow sites can be generated from the triangular complex $Fe_3(CO)_{12}$ by removing the three axial CO ligands on one side of the cluster. This removal generates three adjacent acceptor lobes (Figure 4.13) much like the ones found on the model surface sites of naked clusters as described by Lauher (*393*). A linear combination of these three levels generates three new acceptor orbitals, which are shown near the top of Figure 4.14. The linear combination generates a lower energy $2a_1$ state, in which all the projecting orbital lobes have the same sign, and a $2e$ state. The asymmetric e orbital has two lobes of opposite sign, each centered on a metal atom, whereas the symmetric e orbital has one large lobe of one sign and two tiny lobes of the opposite sign also centered on the metal

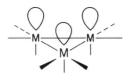

Figure 4.13. Diagram of the threefold site in a metal cluster. The three acceptor lobes shown mix together to form the one a and two e acceptor states of the threefold site.

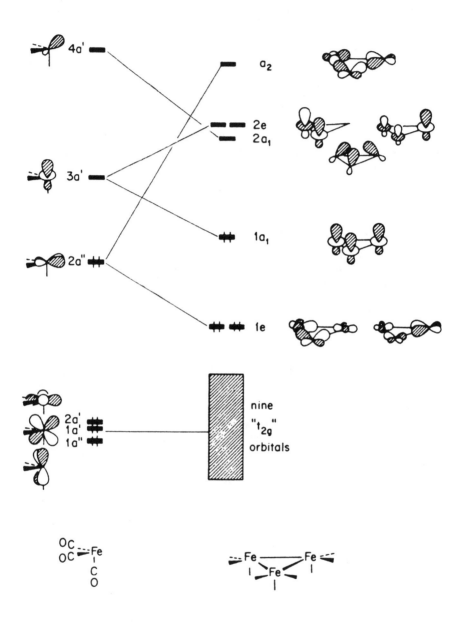

Figure 4.14. Orbital energy diagram of the threefold site (center) generated from the joining of three appropriately oriented $Fe(CO)_3$ fragments (left). The shapes of the resulting orbitals are shown on the right. The new acceptor levels are the 2_{a1} and 2_e states. (Reproduced from reference 43. Copyright 1979 American Chemical Society.)

atoms (*43*). A similar linear combination could be carried out with the three adjacent acceptor lobes calculated to exist on the model (111) site on a naked metal cluster (*393*). This process should generate a qualitatively similar series of acceptor orbitals on a surface threefold site and lead to a similarity of the orbital structure of cluster and surface sites for the threefold hollow site.

As with the other model sites, a similar orbital picture (including the lower filled valence orbitals) can also be generated by joining three $Fe(CO)_3$ fragments that are appropriately aligned to yield the threefold site. This process also locates the next three lower levels of the triangular fragment as well as the high lying a_2 level, as is shown in Figure 4.14 (*43*).

As is the case for atop and bridging sites, the frontier orbital structure of the threefold site can be used to study the bonding of ligands or other fragments in the threefold site. The relative ordering of the levels may, in some cases, change as a function of the electron count of the metal; however, the symmetries and shapes of the levels will remain constant. The kinds of bonding insights that can be gained in the study of these systems provide qualitative insight into surface bonding, even in the absence of band theory.

One of the most well-known ligands to occupy a threefold site is the alkylidyne fragment in the tricobalt nonacarbonyl complex, whose FMO bonding is shown in Figure 4.15 (*43*). When the metal triangle is composed of cobalt atoms, then the $2a_1$ acceptor level is raised in energy, and the highest occupied molecular orbital (HOMO) is the $2e$ level, which contains three electrons. The alkylidyne fragment is *sp* hybridized with a filled *sp* lobe and an *e* state containing one electron composed of two un-rehybridized *p* orbitals on the carbon atom. The bonding interaction involves overlap of the *sp* lobe with the $2a_1$ acceptor level on the fragment. The *e* symmetry *p* orbitals of the alkylidyne fragment overlap with the $2e$ levels on the threefold site and place a total of four electrons in the resulting bonding *e* state (*43*). This interaction is fairly straightforward, even though the hybridization of the alkylidyne fragment is taken as *sp* and not the more commonly assumed sp^3 hybridization. The H atom on the structure in Figure 4.15 can be replaced with any alkyl species. Most likely, the bonding of ethylidyne on a surface would not be much different from the picture described for the tricobalt metal cluster.

A more interesting related species to examine is the carbonium ion of the ethylidyne moiety shown as Structure 4.2 (*43*), which is generated by removing a hydride ion (H^-) from the methyl group of the ethylidyne fragment. The outer carbon atom then becomes sp^2 hybridized and thus has an empty *p* orbital. Unlike ethylidyne, the calculations show that the upright position for this carbonium structure is at an energy maximum. The species tends to bend over so that the empty *p* orbital can overlap with one of the *e* states of the threefold site. If the carbonium ion bends toward a metal atom

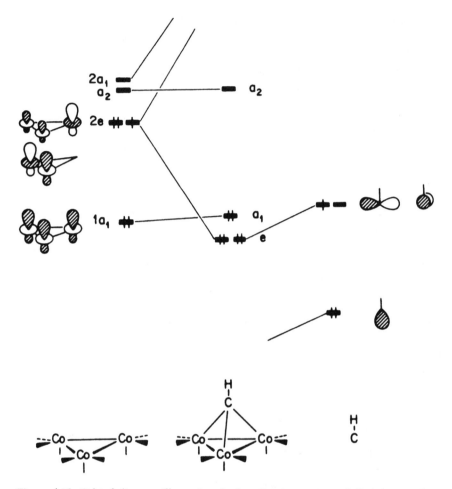

Figure 4.15. Orbital diagram illustrating the bonding interaction of alkylidyne in the threefold site.

Structure 4.2. Illustration of the carbonium ion of the ethylidyne ligand. The "+" sign indicates the presence of an unhybridized p orbital. (Reproduced from reference 43. Copyright 1979 American Chemical Society.)

in the triangle (Figure 4.16a), then the empty p orbital overlaps with the symmetric e state. If it bends toward the center of a metal–metal bond (Figure 4.16), then the empty p orbital will overlap the antisymmetric orbital of the e state. Because the barrier between these two geometries is calculated to be only 16 kcal/mol, the CH_2 moiety could, at ordinary temperature, roll around the triangle about the C–C bond axis.

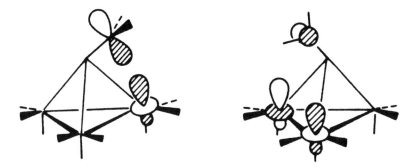

Figure 4.16. Overlaps of the empty p orbital of the carbonium ion with the two e states of the threefold site leading to the tilted structure. (Reproduced from reference 43. Copyright 1979 American Chemical Society.)

Although no such species is known to exist on a surface, the bonding interaction offers understanding of a possible step in the mechanism of preparing surface ethylidyne from acetylene. It has been proposed that a first step in the preparation of ethylidyne is the formation of surface vinylidene $(=C=CH_2)$ (400). What is still unknown, however, is how vinylidene can add an additional hydrogen atom to form the ethylidyne structure. The previous discussion illustrates a possible pathway. When the vinylidene is formed, presumably on a bridging site, it could easily move into the threefold site. This occurrence would lead to not a carbonium ion, but a carbon radical that would probably behave in a manner similar to the carbonium ion because it contains a half-filled p orbital. This species should bend over just as the carbonium ion species does. If the bent structure does form, then the outer carbon atom could easily pick up a surface hydrogen and have the three hydrogen atoms of an ethylidyne moiety. Reworking these calculations for a carbon radical could, therefore, aid our understanding of the mechanism of the acetylene-to-ethylidyne surface reaction.

Another ligand that is known to occupy the threefold site is the sulfur dianion $(S^=)$, a species with three filled p orbitals (43). The interaction diagram for S^{2-} and the $Co_3(CO)_9{}^{2+}$ fragment is shown in Figure 4.17. The interesting aspect of this interaction is that the coordination of the sulfur ligand leads to the population of the a_2 level of the cluster by a single electron. This level lies above the three acceptor orbitals discussed earlier

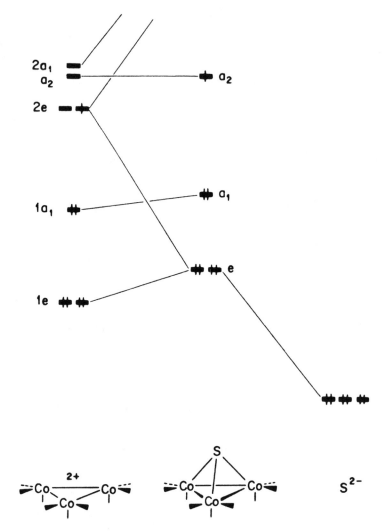

Figure 4.17. Orbital energy diagram of the bonding of the sulfur dianion in a threefold site. The antibonding a_2 level is half-filled. (Reproduced from reference 43. Copyright 1979 American Chemical Society.)

and, as shown in Figure 4.18, is metal–metal antibonding for all three metal–metal bonds in the threefold site. This condition results in an increase of about 0.1 Å in the bond length for each of these metal–metal bonds (43).

By applying this concept to surfaces, it is possible to gain insight into the mechanism of sulfur or oxygen penetration into the bulk of a metal. Both sulfur (401) and oxygen (59) are known to occupy threefold sites on (111) metal surfaces; knowing that their chemisorption bonding may lead to an expansion of their triangular site provides new insight into the nature

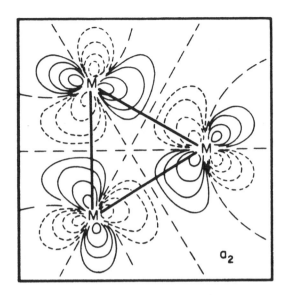

Figure 4.18. Orbital picture of the a_2 orbital of the threefold site showing that the orbital is antibonding for all metal–metal bonds. (Reproduced from reference 43. Copyright 1979 American Chemical Society.)

of the penetration mechanism. This ring expansion may allow the sulfur or oxygen adatom to pop through the surface lattice into the bulk of the metal. Sulfur on a Ni(111) surface causes a significant reconstruction to a two-dimensional surface sulfide (*402*). Perhaps this reconstruction is caused by the sulfur atom expanding its own triangular site and causing a buckling of the Ni(111) surface.

Probably the ligands most relevant to surface science that have been studied by FMO calculations of the threefold site in clusters are the unsaturated hydrocarbons, ethylene and acetylene (*43*). It is possible, by using predominantly symmetry arguments and electron counting, to predict the orientation of the hydrocarbon on the threefold site of the cluster. By assuming that point-group symmetry exists on a surface site, these conclusions may enhance understanding of the bonding geometries of hydrocarbons on surfaces.

For the case of ethylene, symmetry considerations do not allow it to occupy a threefold site in a cluster because the possible orbital overlaps are unfavorable (*43*). This situation is true for geometries where ethylene lies either perpendicular or parallel to one of the metal–metal bonds (Figure 4.19). The filled π-bond level has a destabilizing interaction with both the filled $1a_1$ level and the filled symmetric $1e$ level of the threefold site. The antisymmetric $1e$ level of the threefold site is destabilized by the parallel-lying p-orbital electron density of the ethylene molecule. Because this result is derived solely from symmetry arguments, it raises the question of whether or not ethylene can bond to threefold sites on (111) surfaces (*194*).

Figure 4.19. Two possible geometries for ethylene coordination to a threefold site: ethylene perpendicular to a metal-metal bond (left) and ethylene parallel to a metal–metal bond (right). (Reproduced from reference 43. Copyright 1979 American Chemical Society.)

Acetylene is known to bond on threefold sites in the modes that are both parallel and perpendicular to a metal–metal bond in the triangle. This bonding is determined by the molecular orbital picture of the interaction and the number of electrons in the cluster that can fill the levels (Figure 4.20). The presence or absence of electron density in one level is enough to change the orientation of the acetylene ligand in the threefold site (43).

The essence of the symmetry arguments for the two geometries of the Fe_3 cluster lies in the relative symmetries of the π and π^* levels. Because all of the ligand–metal bonding interactions involve the interactions of three orbitals of the same symmetry—two on the acetylene and one of the two filled e states on the threefold metal site—the orientation and thus the symmetry of the acetylene ligand must be appropriate to allow the threefold interactions to be net bonding. This can occur only if, for a given threefold interaction, one of the acetylene levels is filled and one is empty. This condition will lead to two electrons in the bonding orbital and two electrons in the nonbonding orbital of the three-orbital interaction and result in a situation that is net bonding. When the acetylene is bonded parallel to a metal–metal bond (Figure 4.20, right), both of the filled π levels are symmetric whereas both of the empty π^* levels are antisymmetric. This leads to one net bonding interaction of the two empty π^* levels on the acetylene with the filled orbital in the cluster site and one net repulsive interaction of the two filled π levels on the acetylene with the other filled e orbital in the cluster site. For the geometry in which the acetylene lies perpendicular to a metal–metal bond (Figure 4.20, left), one filled π level is symmetric and one is antisymmetric; the same applies for the empty π^* levels. This condition leads to two net bonding three-level interactions composed of one filled π orbital, one empty π^* orbital, and one of the filled e states on the threefold site, and thus a stable interaction is acheived. The perpendicular geometry is thus the favored geometry for the Fe_3 system (43).

Because there has been no report of the surface structure of acetylene on an Fe(111) surface, this discussion does serve to illustrate the predictive power of symmetry arguments as they are applied to ligand bonding. Similar applications of this type of symmetry argument to surface systems could very likely simplify surface structure characterizations.

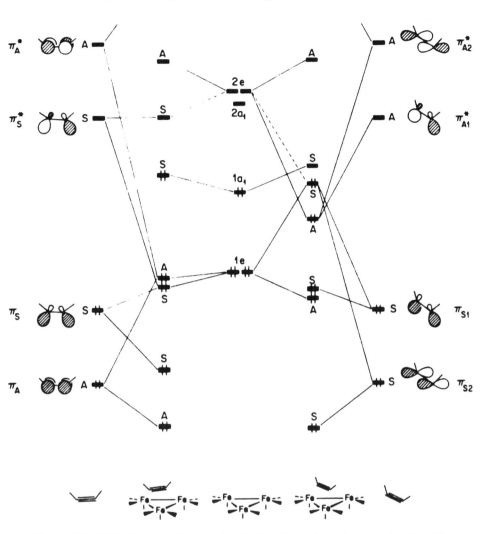

Figure 4.20. Orbital energy diagram for the bonding of acetylene in the threefold site for the two geometries in which the molecule could lie either parallel or perpendicular to a metal–metal bond. The center shows the orbital structure of the threefold site. To either side of the center lie the orbital structures for the two bonding modes. For the parallel geometry (right center), the filled levels are both symmetric (S), whereas the unfilled levels are both antisymmetric (A). This situation leads to one net bonding and one net antibonding interaction and thus a weak acetylene interaction with the threefold site. For the perpendicular geometry (left center), only one acetylene orbital of each symmetry is filled. This situation leads to two net bonding interactions and thus a strong bonding of acetylene to the threefold site. (Reproduced from reference 43. Copyright 1979 American Chemical Society.)

4.3.4   The Long Bridging Site

One of the most interesting bonding sites in organometallic complexes is the long bridging site in which there is no metal–metal bonding interaction (*391b*). It appears in a special class of molecules called A-frame complexes (Figure 4.21). It is important to the surface scientist because it is one type of bonding mode that is not generally considered when characterizing surface chemisorbed species. In the A-frame complex, large chelate ligands hold two metal centers in close proximity (about 3.2 Å apart), yet not close enough for a significant bonding interaction to occur. This model appears to be a very good one for the channel-spanning site on the fcc (110) surface of transition metals (diameter $\cong$ 3.5 Å) and a possible bonding site to be considered when characterizing chemisorption systems on fcc (110) surfaces.

Figure 4.21. Illustration of the long-bridge site in the class of metal complexes called A-frame structures. The metal-metal spacing is too large to allow a significant metal-metal bonding interaction.

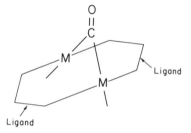

The metal fragment bonding site can be generated by moving two planar ML_3 fragments into close proximity (*391b*). These ML_3 fragments are generated from square planar ML_4 fragments, and each has an acceptor orbital at the location of the missing ligand. This condition is similar to the orbital situation for monometal complex fragments. Because the metal atoms are weakly interacting, it is necessary to consider only the linear combination of these two acceptor levels, which are predominantly of s, p and $d_{x^2-y^2}$ character, in order to arrive at the orbital structure of the acceptor site. The remaining eight d orbitals mix somewhat but remain in a block that does not participate in the bonding at a lower energy.

The acceptor orbital structure of the A-frame bridge site is shown in Figure 4.22. Except for the eight lower levels that are weakly interacting, the site is composed, much like the conventional bridge site, of two orbitals each having two lobes. One of the orbitals has lobes of like parity ($3a_1$), and one has lobes of opposite parity ($3b_2$). Unlike the conventional bridging structure, the orbital with lobes of like parity lies at lower energy. All of the lobes of these two acceptor orbitals point toward the bridging site (*391b*).

The bonding of a ligand in the A-frame bridging site is exemplified by the bonding of the bridging sulfur dianion (Structure **4.3a**) (*391b*). In C_{2v} symmetry, the three filled p orbitals of the sulfur dianon, the p_x, p_y, and p_z,

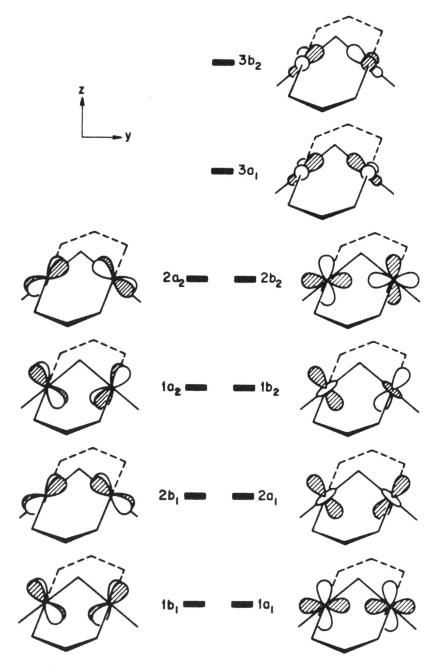

Figure 4.22. Orbital picture of the long-bridge site. The lack of metal-metal bonding is reflected in the molecular orbitals, which are undistorted relative to the atomic orbitals. The top two levels, the 3_{a1} and 3_{b2} compose the acceptor levels of the site. (Reproduced from reference 391b. Copyright 1981 American Chemical Society.)

have symmetries of b_1, b_2, and a_1, respectively. Because of similar symmetries, the p_z can donate electron density to the $3a_1$ acceptor lobe of the
A-frame site, whereas the p_y (of b_2 symmetry) can donate electron density
to the $3b_2$ site (Figure 4.23). The p_x orbital remains essentially nonbonding.
A similar result is obtained for halide anions (Cl⁻, Br⁻, I⁻) except that the
halides are poorer electron donors (*391b*).

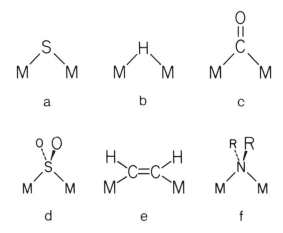

Structures **4.3a-4.3f.** Examples of ligands that are known to occupy the long-bridge
site in A-frame complexes.

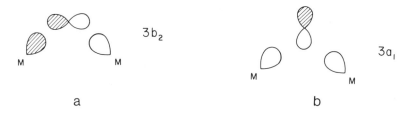

Figure 4.23. Schematic overlap of filled sulfur p orbitals with the acceptor orbitals
of the long-bridge site.

For hydride ligands (H⁻) (Structure **4.3b**), a similar situation arises. The
hydride ligand, however, has only one filled orbital of a_1 symmetry. It
overlaps with the $3a_1$ acceptor lobe on the metal site and leaves the other
acceptor lobe, the $3b_2$ level, empty (*302*).

Because the two possible bonding interactions have a net bonding or
antibonding effect on the interaction of the two metals (*391b*), variations
in the metal–metal distance are observed in A-frame complexes. The a_1
orbital interaction has one sign everywhere between the two metal atoms

and leads to a small metal–metal bonding interaction. The b_2 orbital interaction, because of the presence of the nodal surface in the orbital, leads to a metal–metal antibonding interaction. For the hydride system, therefore, with only the a_1 interaction, a net metal–metal bonding interaction is generated, which leads to a decrease in the metal–metal bond spacing. For the sulfide case, because both levels are filled, no net metal–metal interaction is expected. In fact, the net interaction has been found to be somewhat antibonding, thus leading to a metal–metal spacing greater than that for a nonbonding case (*391b*).

For small-molecule ligands that have a single filled donor orbital and an empty acceptor orbital, the situation is a little different. For a CO bridging ligand (Structure **4.3c**) which has only one donor orbital (the 5σ), it might be expected that CO bonding leads to a decrease in the metal–metal spacing much like the hydride ligand. In fact, this situation is not the case (*391b*); the metal–metal spacing is comparable with that for the sulfur ligand. The reason for this discrepancy is not well understood. By assuming that the CO ligand is formally a ligand with two extra electrons, that is, CO^{2-}, the CO bonding can then be described in a manner much like that for S^{2-}. A similar result is seen for the bridging acetylene species (Structure **4.3e**). Unlike the π-bonded bridging acetylene found in conventional metal–metal bridge sites, acetylene bonds in a di-σ sense on an A-frame bridge site. Even though it only donates a single electron pair, the resulting M–M spacing is more like that for the S^{2-} ligand case. For the SO_2 bridge (Structure **4.3d**), also only a σ donor, a similar result is obtained for the Pd_2 complex; however, a decrease in the metal–metal spacing is detected for the corresponding Rh_2 complex, a species with two fewer electrons (*391b*).

This change in metal–metal spacing may be of significance in surface chemisorption research because it is a possible method of characterizing an adsorbate that bonds in a channel-spanning site on an fcc (110) surface. This change in metal spacing may be reflected in a buckling of the atomic rows along the (001) direction when a channel-spanning adsorbate is present. This buckling may be detectable as a distortion in the shape of the low-energy electron diffraction (LEED) spots or possibly even a splitting of a LEED spot taken from a surface with a channel-spanning adsorbate. Vibrational spectroscopy probably has little application to this problem because the A-frame bridging CO in the complex has a stretching frequency of 1720 cm^{-1}, a value lying within the stretching region of the conventional metal complex bridging CO ligand (*44*). Metal hydride stretches cannot even be detected for A-frame structures (*403*). As yet, there are few reports of channel-spanning surface species on (110) surfaces. A dicarbon channel-spanning species has been reported on Ni(110) (*404*), and it was found that the C_2 fragments occupy every other channel-spanning site. LEED measurements on this system were not reported.

4.4   Benzene Coordination: An Explainable Exception to the Analogy

Although many of the bonding modes for ligands on surfaces qualitatively resemble those in transition metal complexes, there are some surface bonding modes that are noticeably different from modes in metal complexes. However, for at least the case of benzene coordination, it is possible to reconcile the difference in coordination by applying the same orbital arguments discussed in this chapter.

In transition metal complexes, benzene bonds most commonly to monometal centers such as that found in chromocene (Structure **3.53**) (*284*). On the other hand, benzene has been found to bond to the Rh(111) surface (*19*) in threefold sites so that each of the three π-bonds in the ring lies directly over a metal atop site.

These differences in bonding can best be rationalized by first considering the bonding of benzene in chromocene by using the FMO approach (*405*). As shown in Figure 4.24 the coordination site can be generated by removing from the octahedral complex three adjacent ligands in order to form three adjacent acceptor lobes on the complex. This process is similar to the removal of one or two ligands to form monometal acceptor sites as discussed earlier. Linear combination of these levels generates an *a*-symmetry orbital and a pair of orbitals of *e* symmetry. This orbital composition of the bonding site then has three acceptor orbitals of appropriate symmetry to overlap with the three filled levels of the benzene molecule. Overlap of these levels, therefore, leads to the bonding interaction.

Figure 4.24. The three acceptor lobes of the monometal site accommodating benzene in chromocene. These levels mix together to form the one *a* and two *e* states that can overlap with the filled π orbitals of benzene that have the same symmetry.

On surfaces, however, the proposed orbital picture varies greatly from that of chromocene bonding. In chromocene, the metal center, unlike surface metal centers, has no ligands bound to it in the plane perpendicular to the axis of benzene coordination. This condition is what makes the three acceptor lobes available to coordinate the benzene ring. On surfaces, however, a metal center must bond to several atoms in the surface plane and thus can only generate the one acceptor orbital discussed in the section on atop sites on naked metal surfaces. In this case, therefore, the surface can be thought of as being prevented from generating the coordination site most commonly found in organometallic complexes. The system, therefore, must settle for the next best site, which is the threefold site, which has the *a* + *e* orbital pattern shown for the benzene coordination site found in metal complexes. The symmetric distortion found for benzene on Rh(111) (*19*), which places the smaller carbon–carbon bond spacings over the metal centers, could be due to a localization of the three bonding interactions. This situation could indicate a system more similar to the coordination of a ring of three localized, noninteracting π bonds as opposed to a delocalized benzene ring.

4.5   Conclusion

This chapter has illustrated the structural similarity of transition metal bonding sites in metal complexes with those commonly found on transition metal surfaces. The terminal, bridging, and threefold sites that continually appear in discussions of adsorbate bonding on surfaces are seen to be the rule in transition metal complex bonding.

The sites in metal complexes and clusters, however, are unique from surface sites because the metal complex sites can be described in terms of the acceptor and donor orbitals that interact with the ligand coordinating to the site. An understanding of the symmetries of these orbitals, as well as their overlap with the orbitals of the coordinating ligand, has provided a great deal of insight into the nature of the bonding at these sites. Most importantly, these insights derived for metal complexes have been easily extended to issues common to surface chemisorption. Most likely, therefore, theoretical calculations of surfaces that include elements of orbital symmetry within the band structure and elements of orbital overlap can add significantly to the understanding of chemisorption bonding and reactivity on surfaces. The next chapter will illustrate a new theoretical procedure that can achieve these objectives.

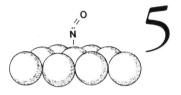

Application of Orbital Symmetry and Orbital Overlap to Surface Band Structure Calculations

IN THE PREVIOUS CHAPTER, frontier molecular orbital calculations of transition metal systems were discussed as a means of illustrating the kinds of insights that can be gained into ligand–metal bonding by considering the specific orbital overlaps that occur when a ligand bonds to a monometal or multimetal site in a transition metal complex. It was shown that the application of symmetry considerations to the metal site, in addition to symmetry considerations of the ligand, serves to clarify several ligand bonding issues. Further, it was pointed out by several examples that similar application of point-group symmetries to surface bonding sites can provide new insights into the surface-adsorbate interaction.

It does not seem intuitively possible to simply ignore issues of symmetry and orbital overlap of surfaces because, on a fundamental level, bonding interactions can occur only when orbitals of the same symmetry overlap. It would seem that this restriction should also apply to surface chemisorption bonds because the symmetries of chemisorption bonds depend upon the adsorbate orbital involved in the bonding. This situation is exemplified in the chemisorption bond of CO, which has both a σ donor bond of a symmetry and a π back bond of e symmetry relative to the CO axis. Although the essentially infinite number of discrete levels in the surface form a continuous band, the individual symmetries of each level must still be somehow manifest so that the σ and π orbitals on the CO molecule can overlap with orbitals of the same symmetry on the surface. It is a fundamental law that orbitals require the same symmetries to overlap, and adsorbate–surface bonding should be no exception.

5.1 The Density of States Projection Calculation

Saillard and Hoffman (24), Sung and Hoffman (25), and, most recently, Silvestre and Hoffman (406a) have addressed this issue in terms of band theory and have devised a method of applying symmetry arguments to surface bonding by deconvoluting or projecting the symmetry and orbital characteristics of the surface band structure from the density of states (DOS) of a solid state system. It is clear from these papers that it is possible to consider adsorbate–surface bonding in first-order terms of orbital overlap and symmetry.

The core of the technique lies in dealing with bunches of levels within the DOS of a surface system rather than one individual level as in a metal complex or cluster (24). When this technique is used, it is possible to extract from the DOS a variety of deconvoluted DOS curves that are important subsets of the total DOS. For instance, Figure 5.1 shows as the dashed line the total DOS for a four-layer Ni(111) slab as calculated by a tight-binding extended Huckel (EH) calculation. This total DOS contains both surface states and bulk states. It is possible, however, to separate the surface DOS from the total DOS and show a deconvolution or projection of the surface DOS only; this projection is shown as the solid line in Figure 5.1 (24). This methodology is important because it is also possible to extract from the total DOS the contribution to the DOS of a given type of metal orbital, that is, the metal s orbital, p_x orbital, d_{z^2} orbital, etc., and thus determine the metal surface DOS of all of those particular orbitals in the metal surface (24).

These DOS projections for individual metal orbitals can be applied to chemisorption bonding in much the same way as individual orbitals in clusters. The DOS projections can be used in understanding the bonding of adsorbed species on surfaces by first calculating the total DOS for a surface covered by an adsorbate and then extracting appropriate DOS projections to determine which levels participate in the bonding. To make this determination, however, it is first necessary to determine the density of states of all the levels that participate in the bonding and antibonding overlaps in the system. This computation is done by calculating what is called a crystal orbital overlap population (COOP) curve, that is, a DOS curve that is weighted by the overlap population (24). The COOP curve is a grouping of either bonding or antibonding states that appear as a series of peaks in a DOS curve. The total DOS is then resolved into projections of the DOS for the individual orbitals; these projections lead to DOS curves that are unique for each orbital type in the system. By comparing the COOP curve with the DOS projection of each individual orbital, peaks in certain of the

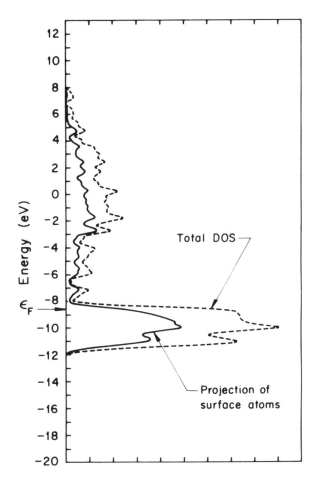

Figure 5.1. Total DOS curve (dashed line) and the deconvolution or projection of the surface DOS (solid line) for a Ni(111) surface. (Reproduced from reference 24. Copyright 1984 American Chemical Society.)

orbital projections will be found to match peaks in the COOP curve. This comparision will lead to a determination of the orbitals that participate in the bonding of the adsorbate to the surface. Even though the DOS projections for each of the orbitals are smeared out over much of the range of the total DOS, there is a noticeable propensity for the DOS of a given orbital to be concentrated in a certain region of the total DOS. This situation simplifies the orbital assignments considerably.

5.2   The Bonding of CO on Ni(100)

The procedure is best illustrated by the example of adsorption of a CO overlayer on Ni(100) (*25*). The total DOS of a CO monolayer on Ni(100)

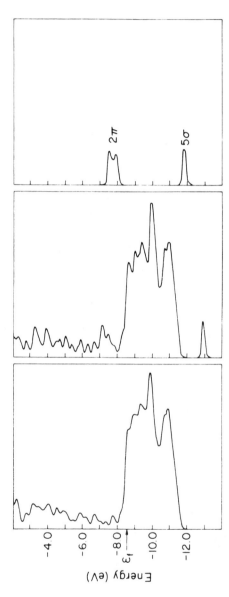

Figure 5.2. Total DOS curves for a Ni(100) slab (left), for a monolayer of CO oriented as a monolayer on a Ni(100) surface (right), and for a monolayer of CO on a Ni(100) surface (center). Note that characteristics of both the CO monolayer and the clean Ni(100) surface can be discerned in the DOS of the combination in the center. (Reproduced from reference 25. Copyright 1985 American Chemical Society.)

is shown in Figure 5.2. The left side of the figure is the total DOS for the Ni(100) surface, whereas the right side is the total DOS for an isolated overlayer of CO arranged in the geometry of a CO overlayer on the Ni(100) surface. When the two systems are brought together, the resulting total DOS becomes that in the center of Figure 5.2. The 5σ band of the CO overlayer is essentially undisturbed, although lower in energy, whereas the 2π band of the CO overlayer is substantially dispersed into the band structure of the Ni(100) surface.

From the total DOS of the system, shown in the center of Figure 5.2, a COOP curve for the M–C bonding interaction can be calculated and is shown as the solid line in Figure 5.3. States that are determined to be metal–carbon bonding appear on the right side of the center line, whereas states that are

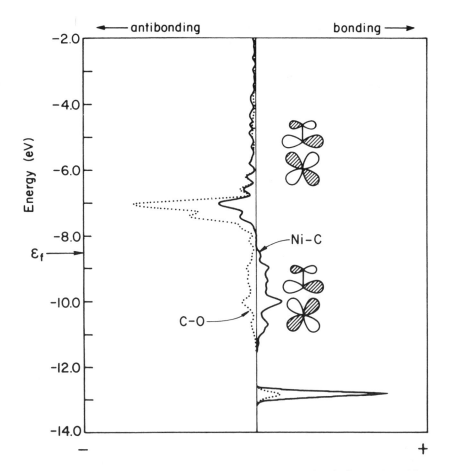

Figure 5.3. COOP curve for the chemisorption of CO on Ni(100). (Reproduced from reference 25. Copyright 1985 American Chemical Society.)

antibonding (M–C or C≡O antibonding) appear on the left side of the center line. All of the bonding orbitals lie below the Fermi level (E_f), whereas all of the antibonding levels appear above the Fermi level. The dotted line of Figure 5.3 is the COOP curve for C–O bonding, which is only antibonding in nature. That the loss of C–O bonding mirrors the appearance of M–C bonding provides insight into the nature of the CO surface bond because the growth of the M–C bond occurs only at the expense of the C–O bonding. This finding is consistent with the decrease in C–O stretching frequency upon chemisorption on nickel.

By taking projections of the total DOS with respect to the orbitals in the system, peaks in the COOP curve can be assigned. From previous experience, it is prudent to look first at the 2π level of CO and the $d\pi$ orbitals of the metal surface, that is, the d_{xz} and d_{yz} orbitals. Figure 5.4 illustrates the projected DOS curves for these two orbital sets. The far left curve shows the 2π DOS for the CO monolayer, whereas the far right curve shows the DOS curve for the $d\pi$ orbitals of the clean Ni(100) surface. The center left curve shows the projection of the DOS for the 2π level in the CO–Ni(100) system, whereas the center right curve shows the DOS of the projection of $d\pi$ levels in the CO–Ni(100) system. From these curves it is clear that in the band structure of the surface and overlayer, although the levels are somewhat smeared out, the DOS of a given level remains predominantly concentrated over a relatively narrow energy range. In addition, the CO

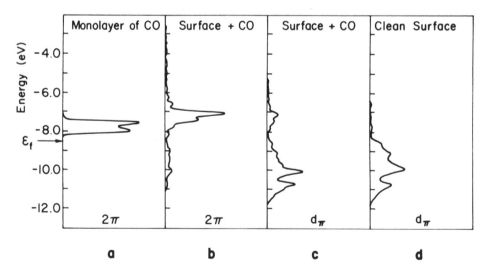

Figure 5.4. DOS projections for (a) the 2π level of a monolayer of CO, (b) the 2π level of adsorbed CO, (c) the $d\pi$ level of the metal surface with an overlayer of adsorbed CO, and (d) the $d\pi$ level of the clean metal surface. The $d\pi$ level refers to metal d orbitals of a point-group symmetry appropriate for back bonding. (Reproduced from reference 25. Copyright 1985 American Chemical Society.)

surface interaction has led to a stabilization of the higher binding energy $d\pi$ bands and a destabilization of the empty low-binding-energy 2π band of CO.

From these observations of the 2π and $d\pi$ DOS projections, the CO–Ni(100) bonding interaction is analogous to the CO bonding in metal complexes. The portion of the COOP curve (Figure 5.3) between -11 and -8 eV can be assigned to $d\pi$ orbitals. Comparison of this region with the DOS projection in the center right of Figure 5.4 shows a general correspondence with regard to the peaks in the two DOS curves. In a similar manner, the peaks between -8 and -6 eV in the COOP curve can be assigned to the 2π bands of CO because the peak shapes of the two DOS regions are similar. Comparison of other DOS projections with the COOP curve in Figure 5.3 does not show a similar correspondence. Therefore, the predominant bonding interaction in the CO/Ni(100) system is an overlap of the 2π DOS of CO with the $d\pi$ DOS of the Ni(100) surface, much like the overlap found in metal complexes.

Even the energy shifts of the 2π band and the $d\pi$ band are consistent with metal complex bonding (Figure 5.5). The downshift of the $d\pi$ level and the upshift of the 2π level lead to the simple orbital overlap picture discussed in Chapter 2. This behavior is qualitatively similar to the back bonding orbital overlap in the bonding of CO to a transition metal center in a complex or cluster. In addition, the lower levels in the COOP curve are M–CO bonding and are predominantly filled; however, the upper levels in the COOP curve are M–CO antibonding, and because they lie above the Fermi energy, they are empty. This situation is also consistent with the back bonding orbital overlap of CO coordination in metal complexes. The only differences lie in the absolute energy values of the levels and the fact that

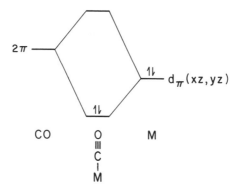

Figure 5.5. Orbital energy diagram for the overlap of the 2π levels of CO with the $d\pi$ levels of a metal atom in metal complexes. Note the qualitative similarities of this picture with that shown for the DOS projections of the corresponding surface orbitals in Figure 5.4.

on surfaces the individual orbitals of a complex tend to be smeared over a small energy range. The concept of orbital symmetry, as discussed earlier for the model surface sites found in clusters, appears to be maintained on a surface for the interaction of CO on Ni(100). This result has been partially validated empirically in a recent study of CO on Ni(111) (*406b*). In this work, surface penning ionization electron spectroscopy (SPIES) was used to observe the partial filling of the 2π level on the CO adsorbate.

The theoretical result just discussed is very much like that for CO bound to naked metal clusters (*39*), which was discussed in the section on frontier molecular orbital theory. As shown for the Fe_3 cluster, the bonding of all the CO ligands is best described by a linear combination of the donor molecular orbitals interacting with a linear combination of the acceptor orbitals on the metal cluster. Although on the cluster these molecular orbitals are delocalized over all the ligands and the metal cluster, it is still appropriate to look upon the CO–metal bonding interaction as localized when considering aspects of reactivity and structure. A similar extension may possibly be made to the surface chemisorption of CO as well as other adsorbates.

5.3 The Bonding of Dihydrogen on Ni(111)

A similar series of arguments has been presented for the interaction of perpendicularly lying dihydrogen with a monometal site on a Ni(111) surface (*24*). This work has shown that although the metal surface is described by band theory, the bonding interaction is qualitatively similar to that found in the transition metal complex. The only difference between the two systems is that, like the CO system, one σ^* orbital of the dihydrogen ligand interacts with one metal d_{xz} orbital in the complex, whereas, on the surface all the dihydrogen σ^* levels in a linear combination interact with all the metal surface orbitals of the appropriate symmetry.

The bonding of the perpendicular dihydrogen species is shown in the COOP curve for the interaction in Figure 5.6. The dashed line represents the orbital overlap for the H–H bond, whereas the solid line represents the overlap contributions to the Ni–H bond. For the H–H bond overlap, the sharp peak at the bottom of the figure represents the σ bonds of the dihydrogen overlayer, which are shown by a DOS projection to disperse only slightly in the band structure. The peak structure at the top of the figure, as seen from a DOS projection, results from the σ^* level of the dihydrogen overlayer. By comparing the peaks in the DOS projections for the individual orbitals with the corresponding peaks in the COOP curve for the M–H bond, we can see how much each contributes to the M–H bond. For the σ level, the M–H COOP curve (solid line) is very small and thus indicates that the σ level participates only to a small extent in the M–H

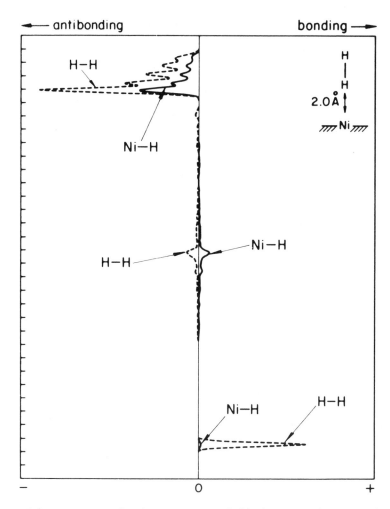

Figure 5.6. COOP curve for the interaction of dihydrogen in the perpendicular geometry with an atop site on a nickel surface. (Reproduced from reference 24. Copyright 1984 American Chemical Society.)

bond. On the other hand, the large peaks of the M-H COOP curve in the σ^* region indicate that the σ^* level participates significantly in the M-H bonding interaction of perpendicular-lying dihydrogen.

The extent and type of participation of the metal bands in the M-H interaction can be determined by the central area of the COOP curve. The fact that the curves for the Ni-H overlap and the H-H overlap mirror each other in this region with regard to bonding and antibonding is an indication that the growth of M-H bonding comes only at the expense of the H-H bonding interaction. This conclusion is reasonable considering the large participation of the σ^* level in the bonding. The orbital contributions can be determined by matching the DOS projections of the various surface orbitals to the peaks in the COOP curve. This COOP curve, however, must have its scale expanded to make the near-Fermi-edge region visible. A blowup of the COOP curve near the Fermi edge is shown in Figure 5.7. DOS projections of the metal s, d_{z^2}, d_{xz}, and d_{yz} levels (Figure 5.8) show that the peaks in Figure 5.7 correspond the closest to peaks in the DOS projections of the s and d_{z^2} orbitals and thus are due predominantly to these levels.

This result compares well with the analogous interaction for the perpendicularly bound dihydrogen ligand with the monometal site in the complex discussed in Chapter 4. In the metal complex, the perpendicularly bound dihydrogen ligand is bound to the metal center through a combination of σ and σ^* orbital overlaps with the large a_1 acceptor orbital of the ML_5

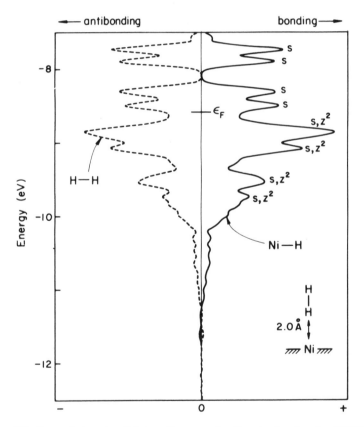

Figure 5.7. Expanded scale of the region near the Fermi edge for the COOP curve shown in Figure 5.6. (Reproduced from reference 24. Copyright 1984 American Chemical Society.)

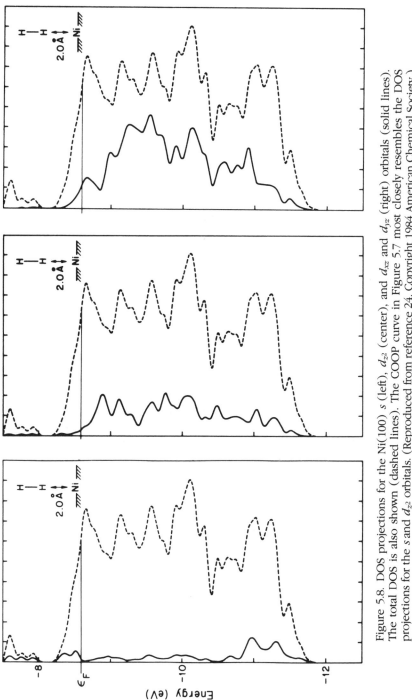

Figure 5.8. DOS projections for the Ni(100) s (left), d_{z^2} (center), and d_{xz} and d_{yz} (right) orbitals (solid lines). The total DOS is also shown (dashed lines). The COOP curve in Figure 5.7 most closely resembles the DOS projections for the s and d_{z^2} orbitals. (Reproduced from reference 24. Copyright 1984 American Chemical Society.)

fragment. Because the complex acceptor orbital is composed of s, p_z, and d_{z^2} metal orbitals, the interaction is qualitatively the same as that calculated for the Ni(111) surface. Quantitatively, the two systems differ primarily because the σ^* is more highly populated for the surface interaction. This finding might be expected, however, because the Ni(111) surface is electron rich (*24*).

Although for the two systems already discussed the symmetries of the interactions on a surface match those of the metal complex, exceptions may arise. The parallel-bound mode for dihydrogen bound to an atop surface site is a case in point. As discussed in Chapter 4, in the complex, the σ level of H_2 overlaps with the a_1 acceptor level of the vacant atop site, whereas the σ^* level overlaps with the filled d_{xz} level of t_{2g} symmetry. On the surface, however, both the σ and σ^* levels are allowed by symmetry to overlap with the t_{2g} levels. Because the σ level is filled, in addition to the t_{2g} levels, the interaction has both a repulsive part as well as an attractive part. The analogy does not completely break down, however, because when the dihydrogen–surface spacing is decreased, the σ^*–d_{xz} overlap increases and leads to a bonding interaction much like that in the metal complex case. The differences in the analogy are reduced to only quantitative elements of the interaction.

5.4   Conclusion

There is, therefore, a significant potential for increasing the understanding of surface chemisorption processes by introducing into theoretical calculations of surfaces aspects of orbital overlap and point-group symmetry as they apply to band theory. The calculations for analogous transition metal complex systems have already illustrated the potential scope of these kinds of investigations. The work involving DOS projections has led to a fairly good qualitative consistency between metal complex and surface calculations and indicates that most likely a similar consistency will be found for the more complicated calculations reviewed in Chapter 4. Surface calculations of this nature that include aspects of symmetry and orbital overlap will undoubtedly be able to enhance the understanding of the structure, bonding, and reactivity of molecules chemisorbed on surfaces.

Literature Cited

1. Muetterties, E. L. *Bull. Soc. Chim. Belg.* **1975**, *84*, 959.
2. Muetterties, E. L. *Bull. Soc. Chim. Belg.* **1976**, *85*, 451.
3. Muetterties, E. L.; Rhodin, T. N.; Band, E.; Brucker, C. F.; Pretzer, W. R. *Chem. Rev.* **1979**, *79*, 91.
4. Schaefer, H. F. III. *Acc. Chem. Res.* **1977**, *10*, 287.
5. Canning, N. D. S.; Madix, R. J. *J. Phys. Chem.* **1984**, *88*, 2437.
6. Muetterties, E. L.; Wexler, R. M. *Surv. Prog. Chem.* **1983**, *10*, 61.
7. Moscovits, M. *Acc. Chem. Res.*, **1979**, *12*, 229.
8. Plummer, W. W.; Salaneck, W. R.; Miller, J. S. *Phys. Rev. B: Solid State* **1978**, *18*, 1673.
9. Conrad, H,,; Ertl, G.; Knozinger, H.; Kuppers, J.; Latta, E. E. *Chem. Phys. Lett.* **1976**, *42*, 115.
10. Plummer, E. W.; Loubriel, G.; Rajoria, D.; Albert, M. R.; Sneddon, L. G.; Salaneck, W. R. *J. Electron Spectrosc. Relat. Phenom.* **1980**, *19*, 35.
11. Albert, M. R. Ph.D Dissertation, University of Pennsylvania, 1983.
12. Costa, N. C. V.; Lloyd, D. R.; Brint, P.; Spalding, T. R.; Pelin, W. K. *Surf. Sci.* **1981**, *107*, L379.
13. Skinner, P.; Howard, M. W.; Oxton, I. A.; Kettle, S. F. A.; Powell, D. B.; Sheppard, N.; *J. Chem. Soc. Faraday Trans. 2*, **1981**, *77*, 1203.
14. Koestner, P. J.; Frost, J. C.; Stair, P. C.; van Hove, M. A.; Somorjai, G. A.; *Surf. Sci.* **1982**, *116*, 85.
15. Demuth, J. E.; Ibach, H. *Surf. Sci.* **1979**, *85*, 365.
16. Avery, N. R. *Appl. Surf. Sci.* **1982**, *14*, 149.
17. Sexton, B. A.; Avery, N. R. *Surf. Sci.* **1983**, *129*, 21.
18. Avery, N. R. *Surf. Sci.* **1984**, *146*, 363.
19. Koel, B. E.; Crowell, J. E.; Mate, C. M.; Somorjai, G. A. *J. Phys. Chem.* **1984**, *88*, 1988.
20. Sappa, E.; Tiripicchio, A.; Braunstein, P. *Coord. Chem. Rev.* **1985**, *65*, 219.

21. Somorjai, G. A. *Chemistry in Two Dimensions*; Cornell University: Ithaca, NY, 1981.
22. Ashcroft, N. W.; Mermin, N. D. *Solid State Physics*; Saunders: Philadelphia, 1976.
23. Kittel, C. *Introduction to Solid State Physics*; Wiley: New York, 1971.
24. Saillard, J. Y.; Hoffman, R. *J. Am. Chem. Soc.* **1984**, *106*, 2006.
25. Sung, S. S.; Hoffman, R. *J. Am. Chem. Soc.* **1985**, *107*, 578.
26. Rudaz, S. L.; Ansermet, J. P.; Wang, P. K.; Slichter, C. P. *Phys. Rev. Lett.* **1985**, *54*, 71.
27. Duncan, T. M.; Yates, J. T., Jr.; Vaughn, R. W. *J. Chem. Phys.* **1980**, *73*, 975.
28. Shibanuma, T.; Matsui, T. *Surf. Sci.* **1985**, *154*, L215.
29. Wang, P. K.; Slichter, C. P.; Sinfelt, J. H. *J. Phys. Chem.* **1985**, *89*, 3606.
30. Duncan, T. M.; Dybowski, C. *Surf. Sci.* **1981**, *1*, 157.
31. van Hove, M. A.; Somorjai, G. *Phys. Rev. Lett.* **1983**, *51*, 778.
32. Kesmodel, L. L.; Dubois, L. H.; Somorjai, G. A. *J. Chem. Phys.* **1979**, *70*, 2180.
33. Prins, R.; Koningsberger, D., Eds. *Principles and Applications of EXAFS, SEXAFS and XANES;* Wiley: NY, 1985.
34. Carver, J. C.; Schweiter, G. K.; Carlson, A. *J. Chem. Phys.* **1972**, *57*, 973.
35. Avanzino, S. C.; Bakke, A. A.; Chen, H. W.; Donohue, C. J.; Jolly, W. L.; Lee, T. H.; Ricco, A. J.; *Inorg. Chem.* **1980**, *19*, 1931.
36. Avanzino, S. C.; Chen, H. W.; Donohue, C. J.; Jolly, W. L. *Inorg. Chem.* **1980**, *19*, 2201.
37. Pauling, L. *General Chemistry*, 3rd ed. Freeman: San Francisco, 1970.
38. Huheey, J. E. *Inorganic Chemistry*; Harper and Row, New York, 1978.
39. Lauher, J. W. *J. Am. Chem. Soc.* **1978**, *100*, 5306.
40. Pauling, L. *The Nature of the Chemical Bond*, 3rd ed. Cornell University: Ithaca, NY, 1960.
41. Burdett, J. K. *Molecular Shapes*; Wiley Interscience: New York, 1980.
42. Cotton, F. A. *Chemical Applications of Group Theory*; Wiley Interscience: New York, 1971.
43. Schilling, B. E. R.; Hoffman, R. *J. Am. Chem. Soc.* **1979**, *101*, 3456.
44. Cotton, F. A.; Wilkinson, G. *Basic Inorganic Chemistry*; John Wiley: New York, 1976.
45. Purcell, K. F.; Kotz, J. C. *Inorganic Chemistry*; Saunders: Philadelphia, 1977.
46. Balahura, R. J.; Lewis, N. A. *Coord. Chem. Rev.* **1976**, *20*, 109.
47. Hitchman, M. A.; Rowbottom, G. L. *Coord. Chem. Rev.* **1982**, *42*, 55.
48. Norbury, A. H. *Adv. Inorg. Chem. Radiochem.* **1975**, *17*, 231.
49. Wexler, R. M.; Tsai, M.-C.; Friend, C. M.; Muetterties, E. L. *J. Am. Chem. Soc.* **1982**, *104*, 2034.
50. Bradley, D. C.; Chisolm, M. H. *Acc. Chem. Res.* **1976**, *9*, 273.
51. Sutton, D. *Chem. Soc. Rev.* **1975**, *4*, 443.

52. Yates, J. T., Jr.; Madey, T. E.; Campuzano, J. C. "The Chemisorption of Carbon Monoxide by the Transition Metals," for *The Physics and Chemistry of Solid Surfaces and Heterogeneous Catalysis,* D. A. King and D. P. Woodruff, Eds., accepted for publication.
53. Loestner, R. J.; van Hove, M. A.; Somorjai, G. A. *J. Phys. Chem.* 1983, *87*, 203.
54. Ittel, S. D.; Ibers, J. A. *Adv. Organomet. Chem.* 1976, *14*, 33.
55. Dewar, M. J. S. *Bull. Soc. Chim. Fr.* 1951, *18*, C79.
56. Chatt, J.; Duncanson, L. A. *J. Chem. Soc.* 1953, 2939.
57. Griffith, W. P. Coord. *Chem. Rev.* 1970, *5*, 459.
58. Griffith, W. P. Coord. *Chem. Rev.* 1972, *8*, 369.
59. Wandelt, K. *Surf. Sci. Rep.* 1982, *2*, 1.
60. Egawa, C., Naito, S., Tamaru, K., *Surf. Sci.* 1983, *125*, 605.
61. Stichney, J. L.; Rosasco, S. D.; Salaita, G. N.; Hubbard, A. T. *Langmuir* 1985, *1*, 66.
62. Renouprez, A. J.; Clughet, G.; Jobic, H. *J. Catal.* 1982, *74*, 296.
63. Bertolini, J. C.; Dalmai-Imelik, G.; Rousseau, J. *Surf. Sci.* 1977, *67*, 478.
64. Sexton, B. A. *Chem. Phys. Lett.* 1979, *65*, 469.
65. Muetterties, E. L. *Chem. Soc. Rev.* 1982, *11*, 283.
66. Bare, S. R.; Stroscio, J. A.; Ho, W. *Surf. Sci.* 1985, *150*, 399.
67. Schoofs, G. R.; Benziger, J. B.; *Surf. Sci.* 1984, *143*, 359.
68. Gates, S. M.; Russell, J. N., Jr.; Yates, J. T., Jr. *J. Catal.* 1985, *92*, 25.
69a. Kingsley, J. R.; Dahlgren, D.; Hemminger, J. C. *Surf. Sci.* 1984, *139*, 417.
69b. Johnson, A. L.; Muetterties, E. L.; Stöhr, J. *J. Am. Chem. Soc.* 1983, *105*, 7183.
70. Hoffman, W.; Bertel, E.; Netzer, F. P. *J. Catal.* 1979, *60*, 316.
71. Gudde, N. J.; Lambert, R. M. *Surf. Sci.* 1983, *124*, 372.
72. George, P. M.; Avery, N. R.; Weinberg, W. H.; Tebbe, F. N. *J. Am. Chem. Soc.* 1983, *105*, 1393.
73. Dorian, P. B.; von Raben, K. U.; Chang, R. K.; Laube, B. L. *Chem. Phys. Lett.* 1981, *84*, 405.
74. von Raben, K. U.; Dorian, P. B.; Chen, T. T.; Chang, R. K. *Chem. Phys. Lett.* 1983, *95*, 269.
75. Knetsch, J. P.; Friedman, H. L. Prog. *Inorg. Chem.* 1983, *30*, 359.
76. Knetsch, D.; Groenwald, W. L.; *Inorg. Chim. Acta* 1973, 7, 81.
77. Nakamoto, K. *Infrared and Raman Spectra of Inorganic and Coordination Compounds,* 3rd edition, Wiley: New York, 1978.
78. Livingstone, S. E. *Coord. Chem. Rev.* 1974, *13*, 101.
79. Verkade, J. G. *Coord. Chem. Rev.* 1972, *9*, 1.
80. Maier, L. *Prog. Inorg. Chem.* 1963, *5*, 27.
81. Weast, R. C., Ed. *CRC Handbook of Chemistry and Physics,* 58th ed. CRC: Cleveland, 1977.
82. King, R. B. *Acc. Chem. Res.* 1980, *13*, 243.

83. Fisher, G. B.; Mitchell, G. E. *J. Electron Spectrosc. Relat. Phenom.* **1983**, *29*, 253.

84. Jacobi, K.; Jensen, E. S.; Rhodin, T. N.; Merrill, R. P. *Surf. Sci.* **1981**, *108*, 397.

85. Klauber, C.; Alvey, M. D.; Yates, J. T., Jr. *Chem. Phys. Lett.* **1984**, *106*, 477.

86. Purtell, R. J.; Merrill, R. P.; Seabury, C. W.; Rhodin, T. N. *Phys. Rev. Lett.* **1980**, *44*, 1279.

87. Sheets, R. W.; Blyholder, G. *J. Catal.* **1981**, *67*, 308.

88. Creighton, J. R.; White, J. M. *Surf. Sci.* **1982**, *122*, 1648.

89. Stockbauer, R.; Hanson, D. M.; Flodstrom, S. A.; Madey, T. E. *Phys. Rev. B: Solid State* **1982**, *26*, 1885.

90. Thiel, P. A.; Hoffman, E. M.; Weinberg, W. H. *Phys. Rev. Lett.* **1982**, *49*, 501.

91. Anderson, S.; Nyberg, C.; Tengstal, C. G. *Chem. Phys. Lett.* **1984**, *104*, 305.

92. Hermann, W. A. *Angew. Chem. Int. Ed. Engl.* **1978**, *17*, 800.

93. Khanra, B. C. *Phys. Lett. A.* **1980**, *76A*, 194.

94. Flores, F.; Gabbay, I.; March, N. H. *Surf. Sci.* **1981**, *107*, 127.

95. Holloway, S.; Benneman, K. H. *Surf. Sci.* **1980**, *101*, 327.

96. Luth, H.; Rubloff, G. W.; Grobman, W. D. *Surf. Sci.* **1977**, *63*, 325.

97a. McBreen, P. H.; Erley, W.; Ibach, H. *Surf. Sci.* **1983**, *133*, 1469.

97b. Paul, J. *Surf. Sci.* **1985**, *160*, 599.

98. Solymosi, F.; Berko, A.; Tarnoczi, T. I. *Surf. Sci.* **1984**, *41*, 533.

99a. Fisher, G. B.; Madey, T. E.; Waclawski, B. J.; Yates, J. T., Jr. Proceedings of the 7th International Vacuum Congress and 3rd International Conference on Solid Surfaces, Vienna 1977; p 1071.

99b. Miles, S. L.; Bernasek, S. L.; Gland, J. L. *J. Phys. Chem.* **1983**, *87*, 1626.

100a. Sexton, B. A.; Hughes, A. E. *Surf. Sci.* **1984**, *140*, 227.

100b. Koestner, R. J.; Stöehr, J.; Gland, J. L.; Kollin, E. B.; Sette, F. *Chem. Phys. Lett.* **1985**, *120*, 285.

101. Friend, C. M.; Muetterties, E. L. *J. Am. Chem. Soc.* **1981**, *103*, 773.

102. Nitschke, F.; Ertl, G.; Kuppers, J. *J. Chem. Phys.* **1981**, *74*, 5911.

103a. Hedge, R. I.; Tobin, J.; White, J. M. *J. Vac. Sci. Technol.* **1985**, *3*, 339.

103b. Hedge, R. I.; White, J. M.; *Surf. Sci.* **1985**, *157*, 17.

104. Hoffman, R.; Chen, M. M.-L.; Thorn, D. L. *Inorg. Chem.* **1977**, *16*, 503.

105. Griffith, W. P. *Coord. Chem. Rev.* **1975**, *17*, 177.

106. Thayer, J. S.; West, R. *Adv. Organomet. Chem.* **1967**, *5*, 169.

107. Corain, B. *Coord. Chem. Rev.* **1982**, *47*, 165.

108. Calabrese, A.; Hayes, R. G. *J. Am. Chem. Soc.* **1974**, *96*, 5054.

109. Solymosi, F.; Kiss, J. *Surf. Sci.* **1981**, *108*, 368.

110. Spencer, N. D.; Lambert, R. M. *Surf. Sci.* **1981**, *104*, 63.

111. Solymosi, F.; Bugyi, L. *Surf. Sci.* **1984**, *147*, 685.

112. DeLouise, L. A.; Winograd, N. *Surf. Sci.* **1985**, *154*, 79.

113. Bauschlicher, C. W. *Surf. Sci.* **1985**, *154*, 70.
114. Goldberg, K. I.; Hoffman, D. M.; Hoffman, R. *Inorg. Chem.* **1982**, *21*, 3863.
115. Allen, A. D.; Harris, R. O.; Loescher, B. R.; Stevens, J. R.; Whitely, R. N. *Chem. Rev.* **1973**, *73*, 11.
116. Tsutsui, M.; Courtney, A. *Adv. Organomet. Chem.* **1977**, *16*, 241.
117. Wang, H. P.; Yates, J. T., Jr. *J. Phys. Chem.* **1984**, *88*, 852.
118. Horn, K.; Dinardo, J.; Eberhardt, W.; Freund, H. J.; Plummer, E. W. *Surf. Sci.* **1982**, *118*, 465.
119. Dowben, P. A.; Sakisaka, Y.; Rhodin, T. N. *Surf. Sci.* **1984**, *147*, 89.
120. Egelhoff, W. F., Jr. *Surf. Sci.* **1984**, *141*, L324.
121. Anton, A. B.; Avery, N. R.; Toby, B. H.; Weinberg, W. H. *J. Electron Spectrosc. Relat. Phenom.* **1983**, *29*, 181.
122a. Miyano, T.; Kamei, K.; Sakisaka, Y.; Ouchi, M. *Surf. Sci.* **1984**, *148*, L645.
122b. Grunze, M.; Golze, M.; Hirschwald, W.; Freund, H.-J.; Plum, H.; Seip, U.; Tsai, M.-C.; Ertl, G.; Kueppers, J. *Phys Rev. Lett.* **1984**, *53*, 850.
123. Ertl, G.; Lee, S. B.; Weiss, M. *Surf. Sci.* **1982**, *114*, 527.
124. Cotton, F. A. *Prog. Inorg. Chem.* **1976**, *21*, 1.
125. Horowitz, C. P.; Shriver, D. F. *Adv. Organomet. Chem.* **1984**, *23*, 219.
126. Eisenberg, R.; Hendriksen, D. E. *Adv. Catal.* **1979**, *28*, 79.
127. Shriver, D. F., Sr.; Alich, A. *Coord. Chem. Rev.* **1972**, *8*, 15.
128. Bonzel, H. P. *J. Vac. Sci. Technol.* **1984**, *A2*, 866.
129. Luftman, H. S.; White, J. M. *Surf. Sci.* **1984**, *139*, 369.
130. Crowell, J. E.; Garfunkel, E. L.; Somorjai, G. A. *Surf. Sci.* **1982**, *121*, 303.
131. Lagow, R. J.; Morrison, J. A. *Adv. Inorg. Chem. Radiochem.* **1980**, *23*, 177.
132. Eberhardt, W.; Hoffman, F. M.; de Paola, R.; Heskett, D.; Stathy, I.; Plummer, E. W.; Moser, H. R. *Phys. Rev. Lett.* **1985**, *54*, 1856.
133. Yates, J. T., Jr.; Kolasinski, K. *J. Chem. Phys.* **1983**, *79*, 1026.
134. Wang, H. P.; Yates, J. T., Jr. *J. Catal.* **1984**, *89*, 79.
135. Eisenberg, R.; Meyer, C. D. *Acc. Chem. Res.* **1975**, *8*, 26.
136. Enemark, J. H.; Feltham, R. D. *Coord. Chem. Rev.* **1974**, *13*, 339.
137. Hoffman, R.; Chen, M. M.-L.; Elian, M.; Rossi, A. R.; Mingos, D. M. P. *Inorg. Chem.* **1974**, *13*, 2666.
138. Hodgson, D. J.; Ibers, J. A. *Inorg. Chem.* **1968**, *7*, 2345.
139. Pierpont, C. G.; Van Deveer, D. G.; Durland, W.; Eisenberg, R. *J. Am. Chem. Soc.* **1970**, *92*, 4760.
140. Frenz, B. A.; Enemark, J. H.; Ibers, J. A. *Inorg. Chem.* **1969**, *8*, 1288.
141. Harrison, B.; Wyatt, M.; Gough, K. G. *Catalysis* **1982**, *5*, 127.
142. Bozso, F.; Arias, J.; Hanrahan, C. P.; Yates, J. T., Jr.; Martin, R. M.; Metiu, H. *Surf. Sci.* **1984**, *141*, 591.
143. Conrad, H.; Scala, R.; Stenzel, W.; Unwin, R. *Surf. Sci.* **1984**, *145*, 1.

144. Wendelken, J. F. *Appl. Surf. Sci.* **1982**, *11/12*, 172.
145. Su, C.-C.; Faller, W. J. *J. Organomet. Chem.* **1975**, *84*, 53.
146. Boca, R. *Coord. Chem. Rev.* **1983**, *50*, 1.
147. Vaska, L. *Acc. Chem. Res.* **1976**, *9*, 175.
148. Gubelman, M. H.; Williams, A. F. *Struct. Bonding* **1983**, *55*, 1.
149. Valentine, J. S. *Chem. Rev.* **1973**, *73*, 235.
150. Shayegan, M.; Cavallo, J. M.; Glover, R. E., IV; Park, R. L. *Phys. Rev. Lett.* **1984**, *53*, 1578.
151. Gland, J. L.; Sexton, B. A.; Fisher, G. B. *Surf. Sci.* **1980**, *95*, 587.
152. Canning, N. D. S.; Outka, D.; Madix, R. J. *Surf. Sci.* **1984**, *141*, 240.
153. Pireaux, J. J.; Chtaib, M.; Delrue, J. P.; Thiry, P. A.; Liehr, T. M.; Cavdano, R. *Surf. Sci.* **1984**, *141*, 211.
154. Sappa, E.; Tiripicchio, A.; Braunstein, P. *Chem. Rev.* **1983**, *83*, 203.
155. Otsura, S.; Nakamura, A. *Adv. Organomet. Chem.* **1976**, *14*, 245.
156. Nast, R. *Coord. Chem. Rev.* **1982**, *47*, 89.
157. Cauletti, C.; Furlani, C.; Piancastelli, M. N.; Sebald, A.; Wrackmeyer, B. *Inorg. Chem.* **1984**, *23*, 1113.
158. Nelson, J. H.; Wheelock, K. S.; Cusachs, L. C.; Jonassen, H. B.; *J. Am. Chem. Soc.* **1969**, *91*, 7005.
159. Hoffman, D. M.; Hoffman, R.; Fisel, R. C. *J. Am. Chem. Soc.* **1982**, *104*, 3858.
160. Fischer, T. E.; Keleman, S. R. *Surf. Sci.* **1978**, *74*, 47.
161. Kesmodel, L. L.; Waddill, G. D.; Gates, J. A. *Surf. Sci.* **1984**, *138*, 464.
162. Stuve, E. M.; Madix, R. J.; Sexton, B. A. *Surf. Sci.* **1982**, *123*, 491.
163. Kiskinova, M.; Pirug, G.; Bonzel, H. P. 8th International Catalysis Conference, Berlin, 1984.
164. Garfunkel, E. L.; Maj, J. J.; Frost, J. C.; Farlas, M. H.; Somorjai, G. A. *J. Phys. Chem.* **1983**, *87*, 3629.
165. Kiskinova, M.; Pirug, G.; Bonzel, H. P. *Surf. Sci.* **1984**, *140*, 1.
166. Netzer, F. P.; Doering, D. L.; Madey, T. E. *Surf. Sci.* **1984**, *143*, L363.
167a. de Paola, R. A.; Hrbek, J.; Hoffman, F. M. *J. Chem. Phys.* **1985**, *82*, 2484.
167b. Lee, J.; Arias, J.; Hanrahan, C. P.; Martin, R. M.; Martin, H. *J. Chem. Phys.* **1984**, *82*, 485.
168a. Butler, I. S. *Acc. Chem. Res.* **1977**, *10*, 359.
168b. Yaneff, P. V. *Coord. Chem. Rev.* **1977**, *23*, 183.
169. Klabunde, K. J.; Kramer, M. P.; Senning, A.; Moltzen, E. K. *J. Am. Chem. Soc.* **1984**, *106*, 263.
170. Hogg, M. A. P.; Spice, J. E. *J. Chem. Soc.* **1958**, 4196.
171. Klabunde, K. J.; White, C. M.; Efner, H. F. *Inorg. Chem.* **1974**, *13*, 1778.
172. Roesky, H. W.; Pandey, K. K. *Adv. Inorg. Chem. Radiochem.* **1983**, *26*, 337.
173. Fowler, A.; Baker, C. J. *Proc. R. Soc. London, Ser. A* **1932**, *136*, 28.
174. Treichel, P. M. *Adv. Organomet. Chem.* **1972**, *11*, 21.
175. Windholz, M., Ed. *The Merck Index*, 9th ed. Merck: Rahway, NJ, 1976.

176. Bland, W. J.; Kemmit, P. D. W.; Moore, R. D.; *J. Chem. Soc. Dalton Trans.* **1973**, *1292.*

177a. Friend, C. M.; Stein, J.; Muetterties, E. L. *J. Am. Chem. Soc.* **1981**, *103*, 767.

177b. Friend, C. M.; Muetterties, E. L.; Gland, J. *J. Phys. Chem.* **1981**, *85*, 3256.

178. Cavanagh, R. R.; Yates, J. T., Jr. *J. Chem. Phys.* **1981**, *75*, 1551.

179. Singleton, E.; Oosthuizen, H. E. *Adv. Organomet. Chem.* **1983**, *22*, 209.

180. Bassett, J. M.; Berry, D. E.; Barker, G. K.; Green, M.; Howard, J. A. K.; Stone, F. G. A. *J. Chem. Soc. Dalton Trans.* **1979**, 1003.

181. Bassett, J. M.; Barker, G. K.; Green, M.; Howard, J. A. K.; Stone, F. G. A. *J. Chem. Soc. Dalton Trans.* **1981**, 219.

182a. Semancik, S.; Haller, G. L.; Yates, J. T., Jr. *J. Chem. Phys.* **1983**, *78*, 6970.

182b. Ceyer, S. T.; Yates, J. T., Jr. *J. Phys. Chem.* **1985**, *89*, 3842.

183. Herman, W. A. *Adv. Organomet. Chem.* **1982**, *20*, 159.

184. Brady, R. C., III; Petit, R. *J. Am. Chem. Soc.* **1980**, *102*, 6181.

185. Kemp, D. S.; Vellachio, F. *Organic Chemistry*; Worth: New York, 1980.

186. Bowden, W. L.; Little, W. F.; Meyer, T. J. *J. Am. Chem. Soc.* **1973**, *95*, 5084.

187. Duckworth, V. F.; Douglas, P. G.; Mason, R., Shaw; B. L. *J. Chem. Soc. Chem. Commun.* **1970**, *70*, 1083.

188. Kloster-Jensen, E. *J. Am. Chem. Soc.* **1969**, *91*, 5673.

189. Gladysz, J. A. *Adv. Organomet. Chem.* **1982**, *20*, 1.

190. Yates, J. T., Jr.; Cavanagh, R. R. *J. Catal.* **1982**, *74*, 97.

191. Wolczanski, P. T.; Bercaw, J. E. *J. Am. Chem. Soc.* **1979**, *101*, 6450.

192. Felter, T. H.; Weinberg, W. H. *Surf. Sci.* **1981**, *103*, 265.

193. Ibach, H.; Hopster, H.; Sexton, B. *Appl. Surf. Sci.* **1977**, *1*, 1.

194. Kesmodel, L. L.; Baetzold, R. C.; Somorjai, G. A.; *Surf. Sci.* **1977**, *66*, 299.

195. Szuromi, P. D.; Engstrom, J. R.; Weinberg, W. H. *J. Chem. Phys.* **1984**, *80*, 508.

196. Ibers, J. A. *J. C. S. Chem. Soc. Rev.* **1982**, *11*, 57.

197. Werner, H. *Coord. Chem. Rev.* **1982**, *43*, 165.

198. Plummer, E. W.; Eberhardt, W. *Adv. Chem. Phys.* **1982**, *49*, 533.

199. Saleh, J. M.; Nasser, F. A. K. *J. Phys. Chem.* **1985**, *89*, 3392.

200. Young, D. A. *Inorg. Chem.* **1973**, *12*, 482.

201. Palmer, D. A.; Eldik, R. V. *Chem. Rev.* **1983**, *83*, 651.

202. Darensbourg, D. J.; Kudaroski, R. A. *Adv. Organomet. Chem.* **1983**, *22*, 129.

203. Sakaki, S.; Kitaura, K.; Morokuma, K. *Inorg. Chem.* **1982**, *21*, 760.

204. Meali, C.; Hoffman, R.; Stockis, A. *Inorg. Chem.* **1984**, *23*, 56.

205. Stuve, E. M.; Madix, R. J.; Sexton, B. A. *Chem. Phys. Lett.* **1982**, *89*, 48.

206. Tanaka, K.; White, J. M. *J. Phys. Chem.* **1982**, *86*, 3977.

207. Dawson, P. H. *Surf. Sci.* **1977**, *65*, 41.

208. Henderson, M. A.; Worley, S. D. *Surf. Sci.* **1985**, *149*, L1.

209. Dawson, P. H. *J. Vac. Sci. Technol.* **1979**, *16*, 1.
210. Bhattacharya, A. K.; Broughton, J. Q.; Perry, D. L. *Surf. Sci.* **1978**, *78*, L689.
211. Dubois, L. H.; Somorjai, G. A. *Surf. Sci.* **1980**, *91*, 514.
212. Dubois, L. H.; Somorjai, G. A. *Surf. Sci.* **1979**, *88*, L13.
213. Solymosi, F.; Kiss, J. *Surf. Sci.* **1985**, *149*, 17.
214. Weinberg, W. H. *Surf. Sci.* **1983**, *128*, L224.
215. Ryan, R. R.; Kubas, G. J.; Moody, D. C.; Eller, P. G. *Struct. Bonding* **1981**, *46*, 47.
216. LaPlaca, S. J.; Ibers, J. A. *Inorg. Chem.* **1966**, *5*, 405.
217. Outka, D. A.; Madix, R. J. *Surf. Sci.* **1984**, *137*, 242.
218. Gainey, T. C.; Hopkins, B. J. *J. Phys. C* **1983**, *16*, 975.
219. Furuyama, M.; Kishi, K.; Ikeda, S. *J. Electron Spectrosc. Relat. Phenom.* **1978**, *13*, 59.
220. Astegger, S. T.; Bechtold, E. *Surf. Sci.* **1982**, *122*, 491.
221. Davies, J. A. *Adv. Inorg. Chem. Radiochem.* **1981**, *24*, 115.
222. Sexton, B. A.; Avery, N. R.; Turney, T. W. *Surf. Sci.* **1983**, *124*, 162.
223. Su, C. C.; Faller, W. J. *Inorg. Chem.* **1974**, *13*, 1734.
224. Roundhill, D. M.; Sperline, R. P.; Beaulieu, W. B. *Coord. Chem. Rev.* **1978**, *26*, 263.
225. Hamilton, L. A.; Landis, P. S.; Kosalapoff, G. M.; Maier, L.; Eds. *Organic Phosphorous Compounds*, Volume 4; Wiley Interscience: New York, 1972.
226. *Chem. Sources - U.S.A.*, 1982 ed; Directories: Ormond Beach, FL, 1982.
227. Hedge, R. I.; Greenlief, C. M.; White, J. M. *J. Phys. Chem.* **1985**, *89*, 2886.
228. Chakravoty, A. *Coord. Chem. Rev.* **1974**, *13*, 1.
229. Jones, R. *Chem. Rev.* **1968**, *68*, 785.
230. Brown, K. L.; Clark, G. R.; Headford, C. E. L.; Marsden, K.; Roper, W. R. *J. Am. Chem. Soc.* **1979**, *101*, 503.
231. Gambarotta, S.; Floriani, C.; Chiesi-Villa, A.; Guastini, C. *J. Am. Chem. Soc.* **1982**, *104*, 2019.
232. Gambarotta, S.; Floriani, C.; Chiesi-Villa, A.; Guastini, C. *J. Am. Chem. Soc.* **1985**, *107*, 2985.
233. Lehwald, S.; Ibach, H.; Demuth, J. E.; *Surf. Sci.* **1978**, *78*, 577.
234. Countrymen, R.; Penfold, B. R. *J. Cryst. Mol. Struct.* **1972**, *2*, 281.
235. Avery, N. R. *Surf. Sci.* **1983**, *125*, 771.
236. Avery, N. R.; Anton, A. B.; Toby, B. H.; Weinberg, W. H. *J. Electron Spectrosc. Relat. Phenom.* **1983**, *29*, 233.
237. Avery, N. R. *Langmuir* **1985**, *1*, 162.
238. Bailey, R. A.; Kozak, S. L.; Michelson, T. W.; Mills, W. N. *Coord. Chem. Rev.* **1971**, *6*, 407.
239. Bottomly, F.; Brooks, W. V. F. *Inorg. Chem.* **1977**, *16*, 501.
240. Avery, N. R. *Surf. Sci.* **1983**, *137*, 501, and references cited therein.

241. Dori, Z. Ziolo, R. F. *Chem. Rev.* **1973**, *73*, 247.
242. Cotton, F. Albert; Wilkinson, G. *Advanced Inorganic Chemistry,* 2nd ed.; Wiley Interscience: New York, 1966.
243. Liu, W., T.; Tsong, T. *Surf. Sci.* **1985**, *151*, 251.
244. Fronaeus, S.; Larsson, R. *Acta Chem. Scand.* **1962**, *16*, 1447.
245. Clark, G. R.; Palenik, G. J. *Inorg. Chem.* **1970**, *9*, 2754.
246. Meek, D. W.; Nicpon, P. E.; Meek, V. I. *J. Am. Chem. Soc.* **1970**, *92*, 5351.
247. Solymosi, F.; Kiss, J. *Surf. Sci.* **1981**, *104*, 181.
248. Solymosi, F.; Rasko, J. *J. Catal.* **1980**, *65*, 235.
249. Solymosi, F.; Berko, A. *Surf. Sci.* **1982**, *122*, 275.
250. Tuan, D. F. T.; Hoffman, R. *Inorg. Chem.* **1985**, *24*, 871.
251. Yu, A.; Tsivadze, A.; Kharitonov, Y. Y.; Tsintsadze, G. V. *Russ. J. Inorg. Chem.* **1972**, *17*, 1417.
252. Solymosi, F.; Bansagi, T. *J. Phys. Chem.* **1979**, *83*, 552.
253. Ashby, R. A.; Werner, R. L. *J. Mol. Spectrosc.* **1965**, *18*, 184.
254. Kiss, J.; Solymosi, F. *Surf. Sci.* **1983**, *135*, 243.
255. Sobota, P.; Janas, Z. *J. Organomet. Chem.* **1984**, *276*, 171.
256. Stamm, W. *J. Org. Chem.* **1965**, *30*, 693.
257a. Surman, M.; Solymosi, F.; Diehl, R. D.; Hoffman, P.; King, D. A. *Surf. Sci.* **1984**, *146*, 144.
257b. Lackey, D.; Surman, M.; King, D. A. *Surf. Sci.* **1985**, *162*, 388.
258. Surman, M.; Solymosi, F.; Diehl, R. D.; Hoffman, P.; King, D. A. *Surf. Sci.* **1984**, *146*, 135.
259. Kim, Y.; Schreifels, J. A.; White, J. M. *Surf. Sci.* **1982**, *114*, 349, and references cited therein.
260. Van Koten, G.; Vrieze, K. *Adv. Organomet. Chem.* **1982**, *21*, 151.
261. Catterick, J.; Thornton, P. *Adv. Inorg. Chem. Nucl. Chem.* **1977**, *20*, 291.
262. Garner, C. D.; Hughes, B. *Adv. Inorg. Chem. Radiochem.* **1975**, *17*, 1.
263. Krishnamurty, K. V.; Harris, G. M. *Chem. Rev.* **1961**, *61*, 213.
264. Burns, R. P.; McAuliffe, C. A. *Adv. Inorg. Chem. Radiochem.* **1979**, *22*, 303.
265. Schrath, W.; Peschel, J. *Chimia* **1964**, *18*, 171.
266. King, R. B. *Inorg. Chem.* **1963**, *2*, 641.
267. Graddon, D. P. *Coord. Chem. Rev.* **1969**, *4*, 1.
268. Livingstone, S. E. *Coord. Chem. Rev.* **1971**, *7*, 59.
269a. Grunze, M. *Surf. Sci.* **1979**, *81*, 603.
269b. Somers, J. S.; Bridge, M. E. *Surf. Sci.* **1985**, *159*, L439.
270. Avery, N. R.; Toby, B. H.; Anton, A. B.; Weinberg, W. H. *Surf. Sci.* **1982**, *122*, L574.
271. Hayden, B. E.; Prince, K.; Woodruff, D. P.; Bradshaw, A. M. *Surf. Sci.* **1983**, *133*, 589.
272. Wendlandt, W. W.; Woodlock, J. H. *J. Inorg. Nucl. Chem.* **1965**, *27*, 259.
273. Adell, B.; Thorlin, G.; *Acta Chem. Scand.* **1950**, *4*, 1.

274. Addison, C. C.; Sutton, D. *Prog. Inorg. Chem.* **1967**, *8*, 195.
275. Krishnamurty, K. V.; Harris, G. M.; Sastri, V. S. *Chem. Rev.* **1970**, *70*, 171.
276. Bradshaw, A. M. *Z. Phys. Chem. Neue Folge.* **1978**, *112*, 33.
277. Horn, R. W.; Weissberger, E.; Collman, J. P. *Inorg. Chem.* **1970**, *9*, 2367.
278. Valentine, J.; Valentine, D., Jr.; Collman, J. P. *Inorg. Chem.* **1971**, *10*, 219.
279. Segner, J.; Vielhaber, W.; Ertl, G. *Isr. J. Chem.* **1982**, *22*, 375.
280. Fuggle, J. C.; Menzel, D. *Surf. Sci.* **1979**, *79*, 1.
281. Kiskinova, M.; Pirug, G.; Bonzel, H. P. *Surf. Sci.* **1984**, *140*, 1.
282. Morrow, B. A.; McFarlane, R. A.; Moran, L. E. *J. Phys. Chem.* **1985**, *89*, 77.
283. Benziger, J. B. *Surf. Sci.* **1984**, *17*, 309.
284. Muetterties, E. L.; Bleeke, J. R.; Wucherer, E. J.; Albright, T. A. *Chem. Rev.* **1982**, *82*, 499.
285. Haaland, A. *Acc. Chem. Res.* **1979**, *12*, 415.
286. Churchill, M. R. *Prog. Inorg. Chem.* **1970**, *11*, 53.
287. Omae, I. *Coord. Chem. Rev.* **1984**, *53*, 261.
288. Rohmer, M. M.; Veillard, A. *Chem. Phys.* **1975**, *11*, 349.
289. Anderson, A. B.; McDevitt, M. R.; Urbach, F. L. *Surf. Sci.* **1984**, *146*, 80.
290. Werner, H. *Adv. Organomet. Chem.* **1981**, *19*, 155.
291. Benn, R.; Rufinska, A.; *Organometallics* **1985**, *4*, 209.
292. Lehwald, S.; Ibach, H.; Demuth, J. E. *Surf. Sci.* **1978**, *78*, 577.
293. Bertolini, J. C.; Rousseau, J. *Surf. Sci.* **1979**, *89*, 467.
294. Lin, R. F.; Loestner, R. J.; van Hove, M. A.; Somorjai, G. A. *Surf. Sci.* **1983**, *134*, 161.
295. Suh, J. S.; Dilella, D. P.; Moskovits, M. *J. Phys. Cem.* **1983**, *87*, 1540.
296a. Mack, J. U.; Bertel, E.; Netzer, F. P. *Surf. Sci.* **1985**, *159*, 265.
296b. Dahlgren, D.; Hemminger, J. C. *Surf. Sci.* **1982**, *114*, 459.
297. Dahlgren, D.; Hemminger, J. C. *Surf. Sci.* **1983**, *134*, 836.
298. Avery, N. R. *Surf. Sci.* **1984**, *137*, L109.
299. Avery, N. R., private communication.
300. Lagowski, J. J. *Coord. Chem. Rev.* **1977**, *22*, 185.
301. Elmes, P. S.; West, B. O. *Coord. Chem. Rev.* **1968**, *3*, 279.
302. Inoue, M.; Kubo, M. *Coord. Chem. Rev.* **1976**, *21*, 1.
303. Holan, D. G.; Hughes, A. N.; Wright, K. *Coord. Chem. Rev.* **1975**, *15*, 239.
304. Netzer, F. P.; Mack, J. U. *Chem. Phys. Lett.* **1983**, *95*, 492.
305. Demuth, J. E.; Christman, K.; Sanda, P. N. *Chem. Phys. Lett.* **1980**, *76*, 201.
306. Chou, C.-C.; Reed, C. E.; Hemminger, J. C.; Ushioda, S.; *J. Electron Spectrosc. Relat. Phenom.* **1983**, *29*, 401.
307. Krasser, W. *J. Mol. Struct.* **1982**, *80*, 187.
308. Richardson, N. V.; Campuzano, J. C. *Vacuum* **1981**, *31*, 449.

309. Unger, S. E.; Cooks, R. G.; Steinmetz, B. J.; Deglass, W. N. *Surf. Sci.* 1982, *116,* L211.
310. Gellman, A. J.; Fariag, M. H.; Salmeron, M.; Somorjai, G. A. *Surf. Sci.* 1984, *136,* 217.
311a. Schoofs, G. R.; Preston, R. E.; Benziger, J. B. *Langmuir* 1985, *1,* 313.
311b. Johnson, A. L.; Muetterties, E. L.; Stöehr, J.; Sette, F. *J. Phys. Chem.* 1985, *89,* 4071.
312. Dahlgren, D.; Heminger, J. C. *Surf. Sci* 1982, *120,* 456.
313a. Moore, D. S.; Robinson, S. D. *Chem. Soc. Rev.* 1983, *12,* 415.
313b. Bau, R.; Teller, R. G.; Kirtley, S. W.; Koetzle, T. F.; *Acc. Chem. Res.* 1979, *12,* 176.
314. McCue, J. P. *Coord. Chem. Rev.* 1973, *10,* 265.
315. Ho, W.; Dinardo, N., J.; Plummer, E. W. *J. Vac. Sci. Technol.* 1980, *17,* 134.
316. Housecroft, C. E.; Fehlner, T. P. *Adv. Organomet. Chem.* 1982, *21,* 57.
317. Grimes, R. N. *Acc. Chem. Res.* 1983, *16,* 22.
318. Grimes, R. N. *Acc. Chem. Res.* 1978, *11,* 420.
319. Bradley, D. C.; Mehrotra, R. C.; Gauer, D. P. *Metal Alkoxides*; Academic: London, 1978.
320. Livingstone, S. E. *Quart. Rev.* 1965, *19,* 386.
321. Gates, J. A.; Kesmodel, L. L. *J. Catal.* 1983, *83,* 437.
322. Ryberg, R. *Chem. Phys. Lett.* 1981, *83,* 423.
323. Erskine, J. L.; Bradshaw, A. M. *Chem. Phys. Lett.* 1980, *72,* 260.
324. Wachs, I. E.; Madix, R. J. *Appl. Surf. Sci.* 1978, *1,* 303.
325. Gates, S. M.; Russell, J. N., Jr.; Yates, J. T., Jr. *Surf. Sci.* 1986, *171,* 111.
326. Russell, J. N., Jr.; Gates, S. M.; Yates, J. T., Jr. *Surf. Sci.* 1985, *163,* 516.
327. Orchin, M.; Schmidt, P. J. *Coord. Chem. Rev.* 1968, *3,* 345.
328a. Mosher, H. S.; Turner, L.; Carlsmith, A. *Org. Synth.* 1953, *33,* 79.
328b. Fukuda, Y.; Rabalais, J. W. *J. Electron Spectrosc. Relat. Phenom.* 1982, *25,* 237.
329a. Gland, J. L.; Fisher, G. B.; Mitchell, G. E. *Chem. Phys. Lett.* 1985, *119,* 89.
329b. Grunze, M.; Brundle, C. R.; Tomanek, D. *Surf. Sci.* 1982, *119,* 133.
330. Davidson, P. J.; Lappert, M. F.; Pearce, R. *Chem. Rev.* 1976, *76,* 219.
331. Parshall, G. W. *Acc. Chem. Res.* 1975, *8,* 113.
332. Wilkinson, G. *Pure Appl. Chem.* 1972, *30,* 627.
333. Seyferth, D. *Adv. Organomet. Chem.* 1976, *14,* 97.
334. Brown, F. J. *Prog. Inorg. Chem.* 1980, *27,* 1.
335. Benziger, J. B.; Madix, R. J. *J. Catal.* 1980, *65,* 49.
336. Steinbach, F. Kiss, J. Krall, R. *Surf. Sci.* 1985, *157,* 401.
337. Wittrig, T. S.; Szuromi, P. D.; Weinberg, W. H. *J. Chem. Phys.* 1982, *76,* 716.
338. Wittrig, T. S.; Szuromi, P. D.; Weinberg, W. H. *J. Chem. Phys.* 1982, *76,* 3305.

339. Lagow, R. J.; Gerchman, L. L.; Jacob, R. A.; Morrison, J. A. *J. Am. Chem. Soc.* **1975**, *97*, 518.

340. Schmidbaur, H. *Acc. Chem. Res.* **1975**, *8*, 62.

341. Kaska, W. C. *Coord. Chem. Rev.* **1983**, *48*, 1.

342. Schmidbaur, H.; Tronich, W. *Chem. Ber.* **1968**, *101*, 595.

343. Triplett, K.; Curtis, M. D. *J. Am. Chem. Soc.* **1975**, *97*, 5747.

344a. Berlowitz, P.; Yang, B. L.; Butt, J. B.; King, H. H. *Surf. Sci.* **1985**, *159*, 540.

344b. McBreen, P. H.; Erley, W.; Ibach, H. *Surf. Sci.* **1984**, *148*, 292.

345. Dubois, L. H.; Castner, D. G.; Somorjai, G. A. *J. Chem. Phys.* **1980**, *72*, 5234.

346. Kesmodel, L. L.; Dubois, L. H.; Somorjai, G. A.; *J. Chem. Phys.* **1979**, *70*, 2180.

347. Albert, M. R.; Sneddon, L. G.; Eberhardt, W.; Gustaffson, T.; Plummer, E. W. *Surf. Sci.* **1982**, *120*, 19.

348. Kesmodel, L. L.; Gates, J. A. *Surf. Sci.* **1981**, *111*, L747.

349. Barteau, M. A.; Broughton, J. Q.; Menzel, D. *Appl. Surf. Sci.* **1984**, *19*, 92.

350. Beebe, T. P., Jr.; Albert, M. R.; Yates, J. T., Jr. *J. Catal.* **1981**, *96*, 1.

351a. Friend, C. M.; Muetterties, E. L. *J. Am. Chem. Soc.* **1981**, *103*, 773.

351b. Beebe, T. P., Jr.; Yates, J. T., Jr. *J. Am. Chem. Soc.* **1986**, *108*, 663.

352. Markby, R.; Wender, I.; Friedel, R. A.; Cotton, F. A.; Sternberg, H. W. *J. Am. Chem. Soc.* **1958**, *80*, 6529.

353. Chesky, P. T.; Hall, M. B. *Inorg. Chem.* **1981**, *20*, 4419, and references cited therein.

354. Mayes, M. J.; Simpson, R. M. F. *J. Chem. Soc. A* **1968**, 1444.

355. Robinson, B. H.; Than, W. S. *J. Chem. Soc. A* **1968**, 1784.

356. Miller, D. C.; Brill, T. B. *Inorg. Chem.* **1978**, *17*, 240.

357. Loestner, R. J.; Frost, J. C.; Stair, P. C.; van Hove, M. A.; Somorjai, G. A. *Surf. Sci.* **1982**, *116*, 85.

358. Beebe, T. P., Jr.; Albert, M. R.; Yates, J. T., Jr., in preparation.

359. Bruce, M. I.; Swincer, A. G. *Adv. Organomet. Chem.* **1983**, *22*, 59.

360. Kesmodel, L. L.; Dubois, L. H.; Somorjai, G. A. *Chem. Phys. Lett.* **1978**, *56*, 267.

361. Davison, A.; Rakita, P. E. *Inorg. Chem.* **1970**, *9*, 289.

362. Campbell, A. J.; Fyfe, C. A.; Goel, R. G.; Maslowsky, E. Jr.; Senoff, C. V. *J. Am. Chem. Soc.* **1972**, *94*, 8387.

363. Baird, M. C. *Prog. Inorg. Chem.* **1968**, *9*, 1.

364. Clark, R. J. H. *Spectrochim. Acta* **1965**, *21A*, 955.

365. Schmid, G. *Agnew. Chem. Int. Ed. Engl.* **1978**, *17*, 392.

366. Wei, C.H.; Dahl, L. F. *Inorg. Chem.* **1967**, *6*, 1229.

367. Vizi-Orosz, A.; Galamb, V.; Palyi, G.; Marko, L. *J. Organomet. Chem.* **1976**, *107*, 235.

368. Vizi-Orosz, A. *J. Organomet. Chem.* **1976**, *111*, 61.

369. Schwarzhans, K. E.; Steiger, H. *Angew. Chem. Int. Ed. Engl.* 1972, *11*, 535.
370. Schmid, G.; Batzel, V.; Etzrodt, G.; Pfeil, R. *J. Organomet. Chem.* 1975, *86*, 257.
371. Schmid, G.; Batzel, V.; Etzrodt, G. *J. Organomet. Chem.* 1976, *112*, 345.
372. Schmid, G.; Etzrodt, G. *J. Organomet. Chem.* 1977, *137*, 367.
373. Hegedus, L. L., McCabe, R. W. *Catalyst Poisoning*; Marcel Dekker: New York, 1984.
374. Trenary, M., Uram, K., Yates, J. T., Jr., *Surf. Sci.* 1985, *157*, 512.
375. Imbihl, R., Behm, R. J., Ertl, G., Moritz, W., *Surf. Sci.* 1980, *123*, 129.
376. Cox, M. P., Lambert, R. M., *Surf. Sci.* 1981, *107*, 547.
377. Spencer, N. D., Lambert, R. M., *Surf. Sci.* 1981, *107*, 237.
378. Erley, W., *Surf. Sci.* 1980, *94*, 281.
379. Avery, N., *J. Chem. Phys.* 1981, *74*, 4202.
380. Bechtold, E., Leonhard, H., *Surf. Sci.* 1985, *151*, 521.
381. Benndorf, C., Kruger, B., *Surf. Sci.* 1985, *151*, 271.
382. Li, C. H., Tang, S. Y., *Phys. Rev. Lett.* 1978, *40*, 46.
383. Plummer, E. Ward, Gustaffson, T., *Science (Washington, D.C.)* 1977, *198*, 165.
384. Muller, A., Jaegermann, W., Enemark, J. H., Coord. *Chem. Rev.* 1982, *46*, 245.
385. Foust, A. S., Foster, M. S., Dahl, L. F., *J. Am. Chem. Soc.* 1969, *91*, 5631.
386. Vizi-Orosz, A., *J. Organomet. Chem.* 1976, *111*, 61.
387. Vaira, M. D.; Sacconi, L., *Angew. Chem. Int. Ed. Engl.* 1982, *21*, 330.
388. Foust, A. S., Foster, M. S., Dahl, L. F., *J. Am. Chem. Soc.* 1969, *91*, 5633.
389. Hoffman, R., *Angew. Chem. Int. Ed. Engl.* 1982, *21*, 711.
390. Elian, M., Hoffman, R., *Inorg. Chem.* 1975, *14*, 1058.
391a. Thorn, D. L., Hoffman, R., *Inorg. Chem.* 1978, *17*, 126.
391b. Hoffman, D. M., Hoffman, R., *Inorg. Chem.* 1981, *20*, 3543.
392a. Shustorovich, E. *J. Phys. Chem.* 1983, *87*, 14.
392b. Shustorovich, E.; Baetzold, R. C. *Science (Washington, DC)* 1985, *227*, 876.
393. Lauher, J. W., *J. Am. Chem. Soc.* 1979, *101* 2604.
394. Mingos, D. M. P., *J. Chem. Soc. Dalton Trans.* 1974, 133.
395. Lauher, J. W., private communication.
396. Kubas, G. J., Ryan, R. R., Swanson, B. I., Vergamini, P. J., Wasserman, H., *J. Am. Chem. Soc.* 1984, *106*, 451.
397. Van Dam, H., Stufkins, D. J., Oskam, A., Doran, M., Hillier, I. H., *J. Electron Spectrosc. Relat. Phenom.* 1980, *21*, 47.
398. Dekock, R. L., Luben, T. V., Hwang, J., Fehlner, T. P., *Inorg. Chem.* 1981, *20*, 1627.
399. Anderson, A. B., *Inorg. Chem.* 1978, *15*, 2598.
400. Kesmodel, L. L., Dubois, L. H., Somorjai, G. A., *J. Chem. Phys.* 1979, *70*, 2180.

401. Marcus, P. M., Demuth, J. E., Jepson, D. W., *Surf. Sci.* **1975**, *53*, 501.
402. Perdereau, M., Oudar, J., *Surf. Sci.* **1970**, *20*, 80.
403. Brown, M. P., Puddephatt, R. J.; Rashidi, R; Seddon, K. R. *J. Chem. Soc. Dalton Trans.* **1978**, *78,* 516.
404. McCarty, J. G.; Madix, R. J., *J. Catal.* **1977**, *48*, 422.
405. Elian, M.; Chen, M. M. L.; Mingos, D. M. P., Hoffman, R., *Inorg. Chem.* **1976**, *15*, 1148.
406a. Silvestre, J.; Hoffman, R.; *Langmuir* **1985**, *1*, 621.
406b. Bozso, F., Yates, J. T., Jr., *J. Chem. Phys.* **1983**, *78*, 4256.

Glossary

Because this book will be read by both organometallic chemists and surface scientists, a glossary is necessary. Many terms used in this book are specific for either organometallic chemistry or surface science, and it is hoped that this glossary will help to familiarize the reader with the words used by the field with which he or she is less familiar. However, each definition can capture at best only the essence of the meaning of a scientific term; for a more complete definition, the reader should consult specialized texts.

In some cases, a term in one of the fields discussed in this book has a term that is roughly synonymous in the other field. These synonymous terms have been placed in italics at the end of the relevant definition. Some have already been suggested by others.

Acetonitrile   Methyl cyanide, $H_3C{\equiv}N{:}$.

Adduct   A general term referring to the product of a reaction in which two or more molecules combine.

Adsorbates   Molecules that are bonded to surfaces. (*ligands*)

Alkali   Elements in Group I of the periodic table: lithium, sodium, potassium, etc.

Alkoxy   Any alcohol that has lost its –OH proton and so has become anionic.

Alkyl group   Any saturated hydrocarbon fragment that has lost a single hydrogen atom. The H atom is usually replaced by another functional group.

Alkylidyne   The general term for the class of hydrocarbon ligands that bond to a threefold site in a metal cluster or on a surface. Example: ethylidyne, $CH_3C{\equiv}$.

Alkyne   Any hydrocarbon containing a carbon–carbon triple bond.

Ambidentate   Ligands that can bond to metal centers through either of at least two different atoms.

1003–9/87/0191$06.00/1 © 1987 American Chemical Society

Anisotropic   Refers to phenomena of magnitudes that change as a function of direction.

Antibonding   A molecular orbital that contains a nodal surface between two bonded atoms in a molecule.

Aromaticity   The added stability a hydrocarbon gains by having $4n + 2$ electrons in its conjugated π orbital structure.

ARUPS   Angle-resolved ultraviolet photoelectron spectroscopy. The study the angular dependence of photoemitted electrons.

Associative adsorption   A molecule attaching itself to a surface without undergoing any bond breakage. (*coordination*)

Axial ligands   Ligands located parallel to the principal symmetry axis of a metal complex.

Back bonding   The electron donation that occurs from the metal atom to the ligand in metal–ligand bonding.

Back donation   Synonym for back bonding.

Band structure   The electronic structure of a solid. The large number of interacting orbitals leads to a continuum of levels. (*molecular orbital diagram*)

Band theory   Theoretical description of the band structure of a solid. (*molecular orbital theory*)

Bent bonded   For a diatomic, X–Y, the bonding mode with a metal center in which the M–X–Y bond angle is less than 180°.

β-Hydride elimination   A decomposition pathway for alkyl ligands bound to metal atoms in which a hydrogen atom on the β-carbon interacts with the metal atom leading to decomposition.

Binding energy   The energy required to remove an electron in a given molecular orbital to infinity.

Bicyclic compound   A molecule in which two rings of atoms share a common bond. The rings are said to be "fused."

Bond order   The number of bonding electron pairs minus the number of antibonding electron pairs associated with a given bond.

Bonding mode   One possible geometry that a ligand can display when bound to a metal center.

Bonding orbital   A molecular orbital that contains no nodal surface between two atoms in a chemical bond.

Bridge bonding   Bonding of a ligand simultaneously to two adjacent metal centers.

Carbene   A hydrocarbon that has lost two hydrogen atoms from one of its carbon atoms. The resulting carbon atom can bond to one or two adjacent metal centers in a complex.

Carbyne   A hydrocarbon that has lost three hydrogen atoms from one of its terminal carbon atoms. The resulting carbon atom can bond to one or three adjacent metal atoms in a cluster.

Channel-spanning site   On an fcc (110) surface, the bridging site that exists between two atoms occupying the adjacent ridges of atoms that lie in the (001) direction. (*long-bridge site*)

Characterization   Refers to the use of any analytical technique to discern the structure of a molecular entity.

Chelate   A ligand that has two or more lone pairs appropriately oriented to occupy individual coordination sites on a metal atom or ion, forming a complex with the ligand.

Chemisorption The formation of a covalent chemical bond between a surface and an adsorbate.

cis A prefix referring to adjacent ligands in a metal complex or compound.

Cluster-surface analogy A term coined by E. Muetterties to refer to the similarities in structure of molecules bound to metal centers in complexes and on surfaces.

Coadsorption The placing on a surface of more than one adsorbate in a single experiment.

Complex A molecule containing one or more metal atoms surrounded by any variety of ligands that donate electron pairs to the metal atoms.

Conjugation A situation in hydrocarbon chains in which alternate C–C bonds are double bonds. Each carbon atom in the chain then has one unhybridized p orbital.

Concerted Describes a chemical reaction in which several bonds are broken and formed simultaneously.

COOP curve Crystal orbital overlap population curve. A density of states weighted by the overlap population for an adsorbed overlayer on a surface.

Coordinate covalent bonding The bonding characterized by the donation of a lone pair by one molecule to another. This type of bonding is found in coordination complexes.

Coordination chemistry The chemistry of metal atoms in which ligands bond to the metal atoms by coordinate covalent bonding.

Coordination complex A molecule containing ligands bound to a cental metal atom by coordinate covalent bonds.

Coordination number The number of nearest neighbors of a given atom.

Coordination site In a metal complex, the location on a metal atom that can accommodate a ligand. (*surface site*)

Coordinative unsaturation The condition when a metal complex has too few ligands to fulfill the 18-electron rule. (*partially covered surfaces*)

Covalent bonding Situation in which each atom participating in a bond donates one electron to the bond.

Crystal field theory A description of ligand bonding in metal complexes that is derived exclusively from electrostatic interactions.

Cumulene A hydrocarbon that has a single carbon atom participating in two double bonds.

CVMO Cluster valence molecular orbital. An orbital in a metal cluster that exerts a bonding influence within the cluster.

Cyanogen Common name for the dimer of the cyanide species: $N{\equiv}C{-}C{\equiv}N$.

Cyclic Term referring to compounds containing rings of atoms.

Dative bonding Synonym for coordinate covalent bonding.

Deconvolution Synonym for projection. The separation of a given subset of orbitals from the total set of orbitals composing the band structure of a solid.

Degenerate Refers to two or more orbitals that have identical energies and symmetries.

Delocalized   Orbitals or electrons that are associated with more than two atoms in a molecule.

Dentate   Refers to the number of sites on a ligand that can coordinate to a metal atom.

Descriptive chemistry   Refers to any branch of chemistry that deals with a variety of individual chemical reactions.

Dewar-Chatt-Duncanson bonding   The bonding mode of olefins and alkynes to metal atoms characterized by the donation of π electron density from the hydrocarbon to the metal atom and back donation of electron density from the metal atom to the π level of the hydrocarbon.

Dialkylamide   A molecular fragment ligand composed of two alkyl groups bound to a nitrogen atom.

Diatomic bending   The hypothetical process in which a terminally bound diatomic passes through the bent, kinked, and flat-lying structures prior to undergoing bond scission.

Diatomic   A molecule containing two atoms. In this book it specifically refers to a class of molecules having a molecular orbital structure similar to that of CO.

Dimer   A molecule containing two identical molecular subunits.

Di-σ bonding   A bonding mode of chemisorbed ethylene or acetylene in which each carbon atom of either adsorbate bonds equivalently to separate surface metal atoms.

Dissociative adsorption   A molecule attaching itself to a surface by undergoing some extent of bond breakage. (*oxidative addition*)

DOS   Density of states. Means of describing the band structure of a solid by determining the distribution of electronic energy levels in the solid as a function of binding energy. (*molecular orbital diagram*)

Dynamical LEED calculations   Calculations that can quantitatively determine the structure of a molecule adsorbed on a single crystal surface from LEED beam intensities as a function of electron energy. (*X-ray crystallography*)

EELS   Electron energy loss spectroscopy. Surface-sensitive technique for determining the vibrational spectrum of an adsorbed overlayer or the phonon spectrum of a solid. (*IR or Raman spectroscopy*)

E_g orbitals   In crystal field theory, the d orbitals of the metal atom $d_{x^2-y^2}$ and d_{x^2} for O_h symmetry) that are destabilized by the electrostatic field generated by an octahedral array of ligands.

18-Electron rule   The requirement that transition metal atoms attain 18 electrons in their valence shell to maximize their stability.

Electron acceptor   A molecule that can enhance its stability by accepting a lone pair of electrons from another molecule.

Electron-deficient bonding   A situation in which three orbitals overlap to form three molecular orbitals that contain a total of only two electrons.

Electron density   A qualitative value describing the total negative charge in a given spatial region of a molecule.

Electron donor   A molecule that can enhance its stability by donating a lone pair of electrons to another molecule or to a surface.

Electronegativity   A relative measure of an atom's tendency to draw electrons to itself.

Electropositive The opposite of electronegative.

ELS Electron loss spectroscopy. A surface spectroscopy in which electron energy losses due to electronic transitions at surfaces are observed.

ESCA Electron spectroscopy for chemical analysis, also known as X-ray photoelectron spectroscopy. The characterization of substances through the determination of the binding energies of their elemental core electrons.

ESDIAD Electron-stimulated desorption ion angular distribution. The study of absorbate structures by exploiting the fact that desorbing adsorbate ionic fragments escape approximately along the axes of their respective bonds.

Exo B-H bond A bond in boron hydride molecules that is not a part of the cage bonding. The boron atom is a part of the cage.

fcc Face-centered cubic.

Fermi edge The highest energy orbital or state in the band structure of a solid that is filled with electrons. (*HOMO*)

Flat lying A surface bonding mode of an adsorbate molecule in which its principal axis is parallel to the surface.

Fluorination The process of replacing hydrogen atoms in a hydrocarbon with fluorine atoms.

Fragment A part of a molecule in which one or more of its atoms do not have a filled valence shell of electrons.

Frontier molecular orbital calculations Calculations in which the orbital structure of molecular fragments are calculated. When the fragments are joined together, their bonding is elucidated by determining how the partially filled orbitals on the two fragments overlap.

Halogeno ligand An anionic halogen: Cl^-, Br^-, or I^-.

Halogens The diatomic molecules found in Group VII of the periodic table: F_2, Cl_2, Br_2, and I_2.

Heterocycles Hydrocarbon molecules that are ring structures that contain atoms other than carbon.

Heteronuclear A molecule containing different atoms.

Heterolefins Molecules having double bonds between different kinds of atoms.

HLAO High-lying atomic orbital. An orbital in a metal cluster that is antibonding in nature and is usually empty because of its higher energy.

HOMO Highest occupied molecular orbital. In a molecule, the lowest binding energy orbital that contains a pair of electrons. (*Fermi edge*)

Homonuclear A diatomic or polyatomic molecule that contains only one type of atom.

HREELS High resolution electron energy loss spectroscopy. Same as EELS.

HSAB Hard–soft acid–base theory. A theory of orbital overlap that includes an added stability factor because of the similarities of the orbitals' sizes.

Hybrid orbitals In valence bond theory, orbitals that result from mixing atomic orbitals (*s, p, d,* etc.) in order to rationalize a molecular structure.

Hydride anion A free hydrogen atom containing two electrons in its $1s$ shell. It carries a negative charge.

Hydrocarbon A class of molecules composed of only hydrogen and carbon atoms.

Hydrogen capping A term coined to refer to the method of delivering anionic molecules to a surface by giving the anion H^+ cations that neutralize the negative charge and makes the molecule volatile.

Hydrogen stripping A term coined to refer to the method of delivering an adsorbate to a surface that contains hydrogen atom(s) that are removed from the molecule upon chemisorption.

Inductive effect The shifting of orbital electron density within a molecule caused by electrostatic fields produced by certain atoms or fragments.

Internuclear axis The line joining the nuclei of two bonded atoms.

Ionic bonding The association of charged molecular species due to electrostatic effects.

Ionization The process of removing an electron from an orbital in a molecule to infinity, most commonly from the HOMO.

IRAS Infrared reflection absorption spectroscopy. The technique that applies infrared spectroscopy to the study of adsorbed overlayers on single crystal surfaces. Resolution of a few wave numbers can be achieved.

Isomers Molecules that contain the same number and type of atoms but have them arranged differently in space.

Jahn-Teller distortion The distortion of a molecule that results in order to lift the degeneracy of a doubly degenerate HOMO state.

Keto-enol tautomerism The tendency of an organic molecule containing an alcohol functionality adjacent to a double bond ($R_2C=C\overset{\displaystyle OH}{\underset{\displaystyle R}{|}}$) to transform that structure into a ketone ($R_2HC-C\overset{\displaystyle O}{\underset{\displaystyle R}{\|}}$) functionality.

Kinetic stability A molecular stability characterized by slow rates of reaction associated with a species' decomposition pathway(s).

L A variable used in molecular structures of coordination compounds. Refers to a ligand in a complex that is not involved in the chemistry being considered.

LEED Low-energy electron diffraction. The method used to determine the unit cell dimensions of an ordered adsorbate overlayer. The method can also determine the unit cell of a clean single crystal surface.

Lewis acid Molecules that are lone electron-pair acceptors.

Lewis base Molecules that are lone electron-pair donors.

Ligand Any molecule that donates a lone pair of electrons to a metal atom in a complex. It may or may not carry a charge.

Ligand field theory A description of ligand bonding in metal complexes that includes aspects of orbital overlap in addition to the electrostatic interactions of crystal field theory.

Linear bonded Diatomic coordination in a metal complex in which both atoms of the diatomic and the metal center are colinear.

Linear combination The mathematical process of adding two or more orbital wave functions together to generate an equal number of molecular orbitals. The process determines how each orbital contributes to the new molecular orbitals.

Linkage isomers Metal complexes of a given metal atom that have the same ligands but differ in the bonding mode of one or more of the ligands.

Lone electron pair Two electrons that occupy a nonbonding orbital in a molecule.

Lone pair donor A molecule with a lone pair of electrons in a low binding energy non-bonding orbital.

Long-bridge site A bridging site between two metal atoms in a complex that interact weakly or not at all. (*channel-spanning site*)

Low-valent complex A metal complex of low oxidation state, that is, zero or less than zero.

LUMO Lowest unoccupied molecular orbital. In a molecule, the highest binding energy orbital that contains no electrons.

Macrocycle A large planar molecule that contains a hole large enough to accommodate a metal atom. Surrounding the hole are lone-pair donor sites that can coordinate to the metal atom.

Main-group elements Atoms that are neither transition metals, lanthanides, or actinides.

Metal cluster A metal complex characterized by each metal atom being bound to at least two other metal atoms.

Methanation The heterogeneous catalytic process of converting CO and hydrogen to methane gas and water.

Methylene bridge A CH_2 ligand bound to two adjacent metal atoms in a complex.

Moiety A part of a molecule that behaves chemically as a unit.

Molecular orbital An orbital in a molecule that results from the combination of atomic orbitals on two or more atoms in the molecule.

Molecular orbital diagram A picture describing the energy levels and degeneracies of the orbitals in a molecule. (*density of states, molecular orbital diagram*)

Molecular orbital theory The description of bonding in molecules that involves the mixing of atomic orbitals to form delocalized orbitals. (*band theory*)

Monodentate A ligand that attaches itself to a metal atom through only one of its atoms.

Monolayer A single layer of adsorbate molecules on a surface that interact directly with the surface.

Monometal complex A metal complex containing only one metal atom.

Multilayer Many layers of adsorbate molecules on a surface. The layer closest to the surface interacts most strongly with the surface. The remaining layers are much like a liquid or solid of the adsorbate.

NEXAFS Near-edge X-ray absorption fine structure. A spectroscopy involving electronic transitions in an adsorbate from filled core levels to empty valence levels. The dependence of the transition probabilities on the angle of polarization of the radiation may be used to determine the structure of the surface species. Also called XANES.

Nitrido A ligand composed of only one nitrogen atom. It carries a -3 charge.

Nitro The name given to the NO_2^- ligand when it coordinates through the nitrogen atom.

Nitrito The name given to the NO_2^- ligand when it coordinates through an oxygen atom.

Nodal surface A mathematical surface within a molecular orbital that has a zero probability of containing the electrons within the orbital.

Nonbonding orbital An orbital that neither increases or decreases the bond order of a bond. Its electron density is generally not associated with a particular bond.

Normal mode A fundamental vibration of a molecule that involves the synchronized movements of all of the atoms in the molecule.

NQR Nuclear quadrupole resonance. A form of nuclear spectroscopy involving transitions between energy levels of a nucleus possessing an electric quadrupole moment as it senses a nonuniform electric field.

Octahedral complex A complex containing six ligands, each lying at a vertex of a regular octahedron.

Octet rule All main group atoms tend to acquire a closed principal shell containing eight electrons.

Olefin Any organic molecule containing a carbon–carbon double bond.

On-top site A surface chemisorption site where the adsorbate molecules bond perpendicularly to one surface metal atom.

Organometallic A branch of chemistry dealing with molecules containing metal–carbon bonds.

Out-of-plane bending Example: The movements of the hydrogen atoms of a planar ethylene molecule when it coordinates to a metal atom in a Dewar–Chatt–Duncanson sense. It indicates rehybridization of the carbon atoms of ethylene.

Overlayer Synonym for monolayer.

Oxidation state A formal assignment of excess charge to an atom in a molecule or a complex, for example, O^2, Ni^{2+}, S^{6+}. The assignment is made on the basis of the relative electronegativities of the atoms and does not correspond to actual ionic charges.

Oxidative addition The addition of two ligands to a metal complex through the breaking of a bond connecting the two ligands. (*dissociative chemisorption*)

Oxo A ligand composed of one oxygen atom. It carries a -2 charge.

Parity Refers to the mathematical sign of a lobe in an atomic or molecular orbital.

Phenyl Refers to a benzene ring that is a part of an organic molecule; $C_6H_5^-$.

ϕ A shorthand notation for the phenyl group, $C_6H_5^-$.

Phosphine PH_3. Also a molecule containing three alkyl groups surrounding a central phosphorus atom.

Phosphite A molecule containing three alkoxy groups surrounding a central phosphorus atom.

Photoemission Synonym for photoelectron spectroscopy.

π bonds Bonds usually formed from the overlap of coplanar p orbitals resulting in an orbital containing two lobes of opposite parity. No electron density of the orbital is found along the internuclear axis of the bond.

Point-group symmetry The symmetry of a molecule as determined by its ability to undergo the fundamental symmetry operations within a given point group.

Polarizability (Raman spectroscopy) A measure of the ability of the electrons in an atom or a molecule to respond to an applied electric field, F, providing an induced dipole, μ ($\mu = \alpha \times F$).

Polycrystalline A solid whose surface consists of a mixture of exposed crystal planes, as is often found in metal foils, metal filaments, and evaporated metal films.

Principal quantum shell All of the orbitals labeled with a given principal quantum number n.

Projection Synonym for deconvolution.

Proximal effect The breakage of bonds to hydrogen atoms through the interaction of the hydrogen atoms with a surface.

Pseudohalogens Molecules that have characteristics like those of the halogens; that is, they have a -1 charge, form strong acids with hydrogen, and dimerize.

R A symbol representing any part of an organic molecule (usually an alkyl group) that does not have a direct, strong influence on the chemistry of the molecule that is being considered.

Reaction coordinate The trajectory on a multicoordinate potential energy surface that best describes a chemical reaction as it proceeds from reactants to products.

Reductive elimination The opposite of oxidative addition. The loss of two ligands from a complex through the bonding together of the two ligands. Two vacant coordination sites are left behind. (*second order desorption*)

Resonance structures In valence bond theory, the different electronic bonding arrangements that can be drawn from the single arrangement of atoms in a molecule. Usually indicates an increased stability relative to the individual electronic structures.

Semibridging CO A recently characterized bonding mode of CO that bridges two metal atoms of differing oxidation states.

SEXAFS Surface extended X-ray adsorption fine structure. A method for surface structure determination involving the observation and analysis of fine structure beyond an X-ray absorption edge; often carried out by the study of the intensity of Auger electrons in order to give surface sensitivity.

σ bond A bond whose electron density lies symmetrically about the internuclear axis of the bond.

Square pyramid A pyramid with a square base. Square pyramidal complexes have five ligands, each pointing to a vertex of this type of pyramid.

Steric A term referring to the inability of two or more fragments of a molecule to occupy the same space at the same time.

Substituent A molecular fragment that is placed on a molecule by a substitution reaction.

Substituted diatomics A class of ligands that have a part resembling a diatomic (such as CO) and part that is organic in nature. Example: CH_3CN.

Surface site The grouping of atoms on a surface that provides a place for an adsorbate molecule to attach itself to the surface. (*coordination site*)

Terminal bonding Ligand bonding to one metal center in a complex or on a surface. (*atop-site bonding*)

Thermodynamic stability The stability of a molecule due to the minimization of its free energy.

Three-center-two-electron bond Synonym for electron-deficient bond. A bond produced by the overlap of three orbitals that contain a total of two electrons. Three molecular orbitals are produced, and only the lowest (bonding) orbital is filled.

TPD Temperature programmed desorption. The study of the gas that desorbs when a surface having an adsorbed overlayer is heated in a programmed fashion.

TPRS Temperature-programmed reaction spectrometry. Similar to TPD, except that a surface reaction occurs during the heating.

trans A prefix used to denote a geometrical structure in which two ligands lie on opposite sides of a metal atom in a complex.

trans **influence** The effect that a ligand has on the characteristics of a ligand trans to it. Several theories have been proposed, but the effect is not well-understood.

Transition metal An element whose atoms have partially filled *d* subshells.

t_{2g} orbitals In crystal field theory, the *d* orbitals of the metal (d_{xz}, d_{xy}, d_{yz}) that are stabilized by the electrostatic field generated by an octahedral array of ligands.

Trigonal bipyramid A solid created by joining two regular tetrahedra through a common face. Trigonal bipyramidal complexes have five ligands, each pointing to one of the vertexes of this solid.

Unsymmetrically bridged CO A bridge bonding mode of CO where the metal carbon bond lengths for one ligand are unequal. Ligands bonding in this manner appear in pairs.

UPS Ultraviolet photoelectron spectroscopy. The study of the valence shell orbital structure of a molecule through the ionization of electrons from the various molecular orbitals of the valence shell. The binding energies of the ionized electrons are determined by using this technique.

Valence bond theory A theory of bonding that involves the overlap of hybrid orbitals on bonded atoms. It cannot predict bond lengths and angles, but it can explain a known structure.

Valence electron An electron occupying an orbital in the outermost principal shell of an atom or molecule.

Valence shell The outermost principal shell of an atom or molecule.

Valence state Synonym for oxidation state.

Vibrational spectroscopy The characterization of molecules or absorbate species by the excitation of their vibrational normal modes.

Wave function The mathematical probability description of electronic charge in an atomic or molecular orbital.

XPS X-ray photoelectron spectroscopy. Synonym for ESCA.

Ylides Molecules in which methyl groups surround certain main-group atoms. Some of the carbon atoms can support a negative charge after losing a hydrogen ion and thus become good ligands for coordination.

Index

203

Editing and indexing by Karen McCeney
Production by Joan C. Cook
Design by Pamela Lewis
Managing Editor: Janet S. Dodd

Typeset by McFarland Company, Dillsburg, PA,
and Hot Type Ltd., Washington, DC
Printed and bound by Maple Press Company, York, PA

Recent ACS Books

Personal Computers for Scientists: A Byte at a Time
By Glenn I. Ouchi
288 pages; clothbound; ISBN 0-8412-1001-2

Writing the Laboratory Notebook
By Howard M. Kanare
145 pages; clothbound; ISBN 0-8412-0906-5

The ACS Style Guide: A Manual for Authors and Editors
Edited by Janet S. Dodd
264 pages; clothbound; ISBN 0-8412-0917-0

Chemical Demonstrations: A Sourcebook for Teachers
By Lee R. Summerlin and James L. Ealy, Jr.
192 pages; spiral bound; ISBN 0-8412-0923-5

Phosphorus Chemistry in Everyday Living, Second Edition
By Arthur D. F. Toy and Edward N. Walsh
342 pages; clothbound; ISBN 0-8412-1002-0

Pharmacokinetics: Processes and Mathematics
By Peter G. Welling
ACS Monograph 185; 290 pages; ISBN 0-8412-0967-7

Phase-Transfer Catalysis: New Chemistry, Catalysts, and Applications
Edited by Charles M. Starks
ACS Symposium Series 326; 195 pages; ISBN 0-8412-1007-1

Geochemical Processes at Mineral Surfaces
Edited by James A. Davis and Kim F. Hayes
ACS Symposium Series 323; 684 pages; ISBN 0-8412-1004-7

Polymeric Materials for Corrosion Control
Edited by Ray A. Dickie and F. Louis Floyd
ACS Symposium Series 322; 384 pages; ISBN 0-8412-0998-7

Porphyrins: Excited States and Dynamics
Edited by Martin Gouterman, Peter M. Rentzepis, and Karl D. Straub
ACS Symposium Series 321; 384 pages; ISBN 0-8412-0997-9

Water-Soluble Polymers: Beauty with Performance
Edited by J. E. Glass
Advances in Chemistry Series 213; 449 pages; ISBN 0-8412-0931-6

Historic Textile and Paper Materials: Conservation and Characterization
Edited by Howard L. Needles and S. Haig Zeronian
Advances in Chemistry Series 212; 464 pages; ISBN 0-8412-0900-6

For further information and a free catalog of ACS books, contact:
American Chemical Society, Sales Office
1155 16th Street N.W., Washington, DC 20036
Telephone 800-424-6747